P9-DGW-241

OSHA COMPLIANCE MANUAL

OSHA COMPLIANCE MANUAL

Joseph M. Roberts, Sr.

RESTON PUBLISHING COMPANY, INC.
A Prentice-Hall Company
RESTON, VIRGINIA

Library of Congress Cataloging in Publication Data

Roberts, Joseph M 1927–

O.S.H.A. compliance manual.

Includes index.

1. Industrial safety—Law and legislation—United States. 2. Industrial hygiene—Law and legislation—United States. I. Title.

KF3570.R6 344'.73'0465 76-790

ISBN 0-87909-599-7

Reston Publishing Company, Inc.

A Prentice-Hall Company

Reston, Virginia 22090

10 9 8 7 6 5 4 3 2 1

PRINTED IN THE UNITED STATES OF AMERICA

CONTENTS

Preface
vii

Cross Index with Occupational Safety and Health Act of 1970
ix

chapter 1
A Basic Explanation of OSHA, the Occupational Safety and Health Act
1

chapter 2
Definitions of Terms
19

chapter 3
Responsibilities and Duties of Employers and Employees
25

chapter 4
Inspections and Investigations
35

chapter 5
Citations and Penalties
47

chapter 6

Enforcement and Right of Appeal
59

chapter 7

Variances
73

chapter 8

Discrimination Protection
81

chapter 9

OSHA Mandatory Accident Reporting
91

chapter 10

OSHA Recordkeeping
95

chapter 11

Mandatory Posting Requirements
123

chapter 12

OSHA Safety and Health Standards
129

chapter 13
Federal Agency Programs
149

chapter 14
Additional Applicable Provisions
217

chapter 15
An Effective Accident Prevention Safety Program
249

chapter 16
Bringing Your Workplace into Compliance
269

Directory for OSHA Offices
279

PREFACE

The Occupational Safety and Health Act, OSHA, was enacted for the purpose of assuring safe and healthful working conditions for every working man and woman in the nation by authorizing enforcement of the standards developed under the act; by assisting and encouraging the states in their efforts to assure safe and healthful working conditions; by providing for research, information, education, and training in the field of occupational safety and health; and for other purposes.

The act applies to every employer and employee in the nation. There are few exceptions. However, each state may take over the enforcement of OSHA in its own state through an approved state plan, which must equal or exceed the federal OSHA program. Check with the OSHA area director for state programs in operation in your area.

Through the enactment of this new legislation, new and stringent occupational safety and health standards have and are being adopted; *compliance is mandatory* for everyone covered by the act.

To effectively carry out the full meaning and intent of the law, an active accident prevention safety program must be established and enacted covering every worker in the nation. Chapter 15 of this manual is a practical, comprehensive accident prevention safety program. In the preceding chapters, a simplified explanation of OSHA is given, and definitions, responsibilities and duties of employers and employees, inspections and investigations, citations and penalties, enforcement and appeals, variances, discrimination, OSHA accident reporting, OSHA recordkeeping, posting requirements, OSHA safety and health standards, and federal agency provisions are explained; numerous additional applicable provisions and guidelines for bringing your workplace into compliance are also provided.

It is the intent of this manual to help make employment safety and health a reality through a realistic accident prevention safety program and a clear understanding of what the OSHA law requires.

Joseph M. Roberts, Sr.

CROSS INDEX

WITH

OCCUPATIONAL SAFETY AND HEALTH ACT

OF 1970

PUBLIC LAW 91--596

Section	*Brief Description*	*Page*
An Act	Be It Enacted	3
2	Basis of the Act	8
2(b)	Stated Purposes of the Act	9
3	Definitions	21
4(a)	Coverage of the Act	4
4(b)(1)	Atomic Energy Act Exemption	13
4(b)(2)	Other Acts Superseded	13
4(b)(3)	Coordination Between Laws	14
4(b)(4)	Workmen's Compensation Law Not Superseded	15
5(a)(1)	Employer to Furnish Safe and Healthful Workplaces	27
5(a)(2)	Employer to Comply with Rules and Regulations	29
5(b)	Employee to Comply with Rules and Regulations	32
6(a)	Promulgation of Standards	139
6(b)	Promulgate, Modify, Revoke Standards	140
6(b)(1)	Advisory Committee Recommendations	140
6(b)(2)	Interested Party Comments and Input	141
6(b)(3)	Objections to Proposed Standards	141
6(b)(4)	Objections to Proposed Standards	142
6(b)(5)	Toxic Material Standards	142
6(b)(6)(A)	Temporary Variance Order	76

Section	*Brief Description*	*Page*
6(b)(6)(B)	Temporary Variance Application	77
6(b)(6)(C)	Variance for Experimental Work	78
6(b)(7)	Warnings, Labels, Protective Equipment	143
6(b)(8)	Differences with National Consensus	144
6(c)	Temporary Standards	145
6(d)	Permanent Variances	79
6(e)	Publication Requirements	145
6(f)	Petition for Judicial Review	146
6(g)	Priority for Establishing Standards	146
7(a)	National Advisory Committee	219
7(b)	Standard-Setting Advisory Committee	220
7(c)	Assistance for the Secretary of Labor	221
8(a)	Right to Inspect and Investigate	37
8(b)	Subpoena Power	42
8(c)(1)	Provision for Recordkeeping and Posting Notices	103 & 125
8(c)(2)	Provision for Reporting and Recordkeeping	94 & 103
8(c)(3)	Recordkeeping for Toxic Materials, etc.	104
8(d)	Minimum Burden Provision	105
8(e)	Right to Accompany Inspector	42
8(f)	Employee's Right to Request Inspection	43
8(g)(1)	Publication of Reports	44
8(g)(2)	Rules for Inspection	45
9(a)	Citations Issued	49
9(b)	Citations Posted	50
9(c)	Citations Time Limitation	50
10(a)	Notification of Penalty	63
10(b)	Notification of Failure to Correct	64
10(c)	Employer's Notification to Contest	65
11(a)	Judicial Review	66
11(b)	Judicial Review	68
11(c)(1)	Discrimination Prohibited	84
11(c)(2)	Employee's Discrimination Complaint	85
11(c)(3)	Notification of Determination	86
12	OSHA Review Commission	222

Section	*Brief Description*	*Page*
13	Imminent Dangers	70
14	Representation in Civil Litigation	71
15	Protection of Trade Secrets	45
16	Variations, Tolerances, and Exemptions	79
17(a)	Repeated Violation—Penalty	51
17(b)	Serious Violation—Penalty	52
17(c)	Nonserious Violation—Penalty	54
17(d)	Failure to Correct—Penalty	54
17(e)	Willful Violation—Penalty	55
17(f)	Advance Notice—Penalty	55
17(g)	Falsification—Penalty	56
17(h)	Killing Enforcement Personnel—Penalty	56
17(i)	Violation of Posting Requirements—Penalty	57
17(j)	Authority to Assess Penalties	57
17(k)	Serious Violation Definition	24 & 52
17(l)	Payment of Penalties	58
18	State Jurisdiction and State Plans	225
19	Federal Agency Programs	149
20	Research and Related Activities	228
21	Training and Employee Education	231
22	National Institute	232
23	Grants to States	235
24	Statistics	236
25	Audits	238
26	Agency's Annual Report	239
27	National Commission Workmen's Compensation Laws	240
28	Aid to Small Businesses	244
29	Additional Assistant Secretary of Labor	245
30	Additional Positions	246
31	Emergency Locator Beacons	246
32	Separability	247
33	Appropriations	248
34	Effective Date	248
	Legislative History	248

I

A BASIC EXPLANATION OF OSHA, THE OCCUPATIONAL SAFETY AND HEALTH ACT

A BASIC EXPLANATION OF OSHA, THE OCCUPATIONAL SAFETY AND HEALTH ACT

WHAT IS OSHA?

OSHA is the Occupational Safety and Health Act of 1970. This federal law, which took effect April 28, 1971, requires mandatory compliance by every employer and employee in the nation (with few exceptions), and is designed to assure safe and healthful working conditions for every worker in the nation.

"AS STATED"
AN ACT
To assure safe and healthful working conditions for working men and women; by authorizing enforcement of the standards developed under the Act; by assisting and encouraging the States in their efforts to assure safe and healthful working conditions; by providing for research, information, education, and training in the field of occupational safety and health; and for other purposes.

Be it enacted by the Senate and House of Representatives of the United States of America in Congress assembled, That this Act may be cited as the "Occupational Safety and Health Act of 1970." Public Law 91-596, 91st Congress, S. 2193, December 29, 1970.

DOES THIS LAW APPLY TO YOU?

Yes, with few exceptions. You must comply if you are an employer or an employee in any of the following locations:

Any state
District of Columbia
Trust Territory of the Pacific Islands

Puerto Rico
Virgin Islands
American Samoa
Guam
Wake Island
Outer Continental Shelf lands
Johnston Island
Canal Zone

"AS STATED"
Sec. 4. "THE ACT"
Sec. 4. (a) *This Act shall apply with respect to employment performed in a workplace in a State, the District of Columbia, the Commonwealth of Puerto Rico, the Virgin Islands, American Samoa, Guam, the Trust Territory of the Pacific Islands, Wake Island, Outer Continental Shelf lands defined in the Outer Continental Shelf Lands Act, Johnston Island, and the Canal Zone. The Secretary of the Interior shall, by regulation, provide for judicial enforcement of this Act by the courts established for areas in which there are no United States district courts having jurisdiction.*

In general, any employer employing one or more employees who is engaged in a business that in any way affects interstate commerce is covered by this workplace safety and health law.

Because of the broad scope of the act, some employers will be covered who have been exempt from many or all of the other federal labor laws. Some examples are the following:

The professions: Attorneys, physicians, and members of other professions who have one or more employees, even a private secretary, are covered.

Agricultural employers: Anyone engaged in agricultural activity—cultivation, canning, etc.—who has one or more employees is covered. However, members of an agricultural employer's immediate family are not considered "employees."

Nonprofit organizations: Any charitable or nonprofit organization that employs one or more employees is covered, for example, disaster relief organizations, philanthropic organizations, trade associations, private educational institutions, labor organizations, and voluntary hospitals.

Religious organizations: Religious groups are considered employers under the act if they employ one or more persons in secular activities. However, performance of, or participation in, religious services is not considered employment.

Special rule for manpower trainees: Many persons are receiving occupation or job training under Labor Department contracts, but are not employees of the contractor. Therefore, they are not directly protected by the act, which refers only to employees.

However, the secretary of labor has issued regulations requiring all such contractors to include in their contracts an agreement to give these trainees the same safety and health protection as provided by the act, keep the same records, and include these clauses in all subcontracts. Failure to comply subjects the contractor to loss of his contract and debarment from future contracts.

This means, for example, that a trainee being taught bulldozer operation by a company under the Labor Department's Job Corps program will be afforded the same protection as employees of that company.

IS THERE ONLY ONE OSHA LAW WITH WHICH YOU MUST COMPLY?

The *federal* Occupational Safety and Health Act of 1970 governs overall OSHA programs. However, under provisions in the federal OSHA law, each state may take over the OSHA programs in its own individual state. (See Section 18 of the act, which is cross-indexed at the front of this manual.)

State Plans

States operating under approved state plans that allow them to enforce their own safety and health standards may have even tighter coverage requirements. If you are in one of those states, be sure to check your state plan to see if you are covered.

State OSHA programs must be approved and are continuously monitored by the federal OSHA Office to assure that the state programs equal or exceed the federal OSHA programs. At any time the Labor Department finds a state is not complying with the conditions for approval of a state plan, it can withdraw approval. Then the state plan will no longer be in effect, and you will have to comply with all federal requirements.

Check with the OSHA area director (see the Directory at the back of this manual) for the approved state plan in effect in your state.

WHERE DOES THE FEDERAL OSHA APPLY? AND WHERE DO STATE OSHA PROGRAMS APPLY?

Where there is *no* approved *state* OSHA program, you must comply with the federal OSHA requirements. Where there *is* an approved *state* OSHA program, you must comply with the state OSHA requirements.

HOW TO BEST UNDERSTAND WHAT YOU MUST COMPLY WITH

The problem of OSHA compliance can be separated into three basic areas of concern or involvement as follows:

1 / Understanding the law and its administrative regulations.

2 / OSHA Safety and Health Standards.

3 / Employer–employee participation.

First: You must understand the law and the administrative regulations with which you must comply. This manual covers every section of the law, including the administrative regulations (i.e., accident reporting, recordkeeping, posting, citations, appeals, variances, etc.). Some administrative regulations require you to continuously maintain records on prescribed forms, report accidents, and keep employees notified of their rights. Other administrative regulations only involve your compliance when dealing with a specific subject, i.e., appeals, variance applications, etc.

Second: You must comply with OSHA Safety and Health Standards. Chapter 12 of this manual deals with these standards.

Third: Understanding the law, administrative regulations, and standards only puts you in a position of knowing. Chapter 15 of this manual provides you with possibly the most important ingredient to assure your compliance with this mandatory law, a basic accident prevention safety program. Employer–employee participation can best be assured through an organized and effective accident prevention safety program.

CAN YOU GET ASSISTANCE?

Yes! The law gives you three breaks:

1 / If you qualify under the Small Business Act, you can apply for a small business loan to help you meet the cost of complying.

2 / If required medical examinations are for research purposes, e.g., to determine whether employees are adversely affected by a certain hazard, you may be able to get the government to pay for them.

3 / You can ask the National Institute for Occupational Safety and Health (NIOSH) to determine whether any substance in your workplace is potentially toxic. If NIOSH finds that your employees are exposed to a toxic substance, they will tell you how to eliminate the hazard without triggering a full-scale inspection.

SHOULD YOU ASK FOR OUTSIDE HELP?

Yes! Do not make the mistake of thinking that you know all about how to keep your workplace safe. Get help. Contact your insurance carrier for ideas, subscribe to safety publications, contact your employers association to coordinate safety programs, and consider sending key people to OSHA seminars in your area.

Try a NIOSH "walk through," which is a free consultation "inspection" of your workplace. Since NIOSH has no enforcement powers, you will not get any citations or penalties if NIOSH personnel find violations. A survey team consisting of experts in industrial hygiene, occupational safety, and occupational medicine will come to your workplace. They will conduct an in-depth "walk through" of your plant, keyed specifically to OSHA standards. At the end of the "inspection," they will summarize any violations they have found and make recommendations on how to eliminate them before you are hit with OSHA citations.

WHAT IS THE BASIS FOR THE OSHA LAW?

Concern over occupational (job-related) accidents and illnesses has resulted in far-reaching legislation designed to focus attention on the cause of the problem, as well as to provide standards for preventing future occurrences. Congress has determined that a substantial burden is being imposed upon others due to work injuries and illnesses. For example, the OSHA secretary has said that "the loss of a single employee, whether from a short-term illness or injury or a permanent loss from death, can spell disaster to a small business." He noted that a mail-order house would have to fill 1,087 two-dollar orders to pay the cost of a single $100 accident. A department store would have to sell 5,128 pairs of boys' socks, a bakery would have to produce 1,351 loaves of bread, and a restaurant would have to serve 1,351 two-dollar lunches to cover the cost of that single $100 accident.

> "AS STATED"
> Sec. 2. "THE ACT"
> Sec. 2. *The Congress finds that personal injuries and illnesses arising out of work situations impose a substantial burden upon, and are a hindrance to, interstate commerce in terms of lost production, wage loss, medical expenses, and disability compensation payments.*

WHAT IS THE STATED PURPOSE OF THE OSHA LAW?

The stated purpose is to assure safe and healthful working conditions for every worker in the nation.

"AS STATED"
Sec. 2. "THE ACT"
Sec. 2.(b) *The Congress declares it to be its purpose and policy, through the exercise of its powers to regulate commerce among the several States and with foreign nations and to provide for the general welfare, to assure so far as possible every working man and woman in the Nation safe and healthful working conditions and to preserve our human resources.*

HOW ARE THE GOALS OF OSHA TO BE MET?

1 / By employers and employees reducing hazards and instituting safety programs.

"AS STATED"
Sec. 2. "THE ACT"
Sec. 2.(b)(1) *by encouraging employers and employees in their efforts to reduce the number of occupational safety and health hazards at their places of employment, and to stimulate employers and employees to institute new and to perfect existing programs for providing safe and healthful working conditions;*

2 / By employers and employees having separate but dependent responsibilities and rights.

"AS STATED"
Sec. 2. "THE ACT"
Sec. 2.(b)(2) *by providing that employers and employees have separate but dependent responsibilities and rights with respect to achieving safe and healthful working conditions;*

3 / By mandatory safety and health standards.

"AS STATED"
Sec. 2. "THE ACT"

Sec. 2.(b)(3) *by authorizing the Secretary of Labor to set mandatory occupational safety and health standards applicable to businesses affecting interstate commerce, and by creating an Occupational Safety and Health Review Commission for carrying out adjudicatory functions under the Act;*

4 / By building on advances already made.

"AS STATED"
Sec. 2. "THE ACT"
Sec. 2.(b)(4) *by building upon advances already made through employer and employee initiative for providing safe and healthful working conditions;*

5 / By providing for research and development.

"AS STATED"
Sec. 2. "THE ACT"
Sec. 2.(b)(5) *by providing for research in the field of occupational safety and health, including the psychological factors involved, and by developing innovative methods, techniques, and approaches for dealing with occupational safety and health problems;*

6 / By exploring ways to discover and eliminate latent diseases and health problems.

"AS STATED"
Sec. 2. "THE ACT"
Sec. 2.(b)(6) *by exploring ways to discover latent diseases, establishing causal connections between diseases and work in environmental conditions, and conducting other research relating to health problems, in recognition of the fact that occupational health standards present problems often different from those involved in occupational safety;*

7 / By providing that no employee will suffer diminished health, functional capacity, or life expectancy.

"AS STATED"
Sec. 2. "THE ACT"
Sec. 2.(b)(7) *by providing medical criteria which will assure insofar as practicable that no employee will suffer diminished health, functional capacity, or life expectancy as a result of his work experience;*

8 / By providing for training.

"AS STATED"
Sec. 2. "THE ACT"
Sec. 2.(b)(8) *by providing for training programs to increase the number and competence of personnel engaged in the field of occupational safety and health;*

9 / By providing for development and establishing safety and health standards.

"AS STATED"
Sec. 2. "THE ACT"
Sec. 2.(b)(9) *by providing for the development and promulgation of occupational safety and health standards;*

10 / By an effective enforcement program.

"AS STATED"
Sec. 2. "THE ACT"
Sec. 2.(b)(10) *by providing an effective enforcement program which shall include a prohibition against giving advance notice of any inspection and sanctions for any individual violating this prohibition;*

11 / By encouraging state participation.

"AS STATED"
Sec. 2. "THE ACT"
Sec. 2.(b)(11) *by encouraging the States to assume the fullest responsibility for the administration and enforcement of their occupational safety and health laws by providing grants to the States to assist in identifying their needs and responsibilities in the area of occupational safety and health, to develop plans in accordance with the provisions of this Act, to improve the administration and enforcement of State occupational safety and health laws, and to conduct experimental and demonstration projects in connection therewith;*

12 / By requiring reporting and recordkeeping.

"AS STATED"
Sec. 2. "THE ACT"
Sec. 2.(b)(12) *by providing for appropriate reporting procedures with respect to occupational safety and health which procedures will help achieve the objectives of this Act and accurately describe the nature of the occupational safety and health problem;*

13 / Through joint labor and management efforts.

"AS STATED"
Sec. 2. "THE ACT"
Sec. 2.(b)(13) *by encouraging joint labor–management efforts to reduce injuries and disease arising out of employment.*

ATOMIC ENERGY ACT EXEMPTION

Work being performed under the Atomic Energy Act of 1954 (as amended) is exempt from the OSHA law.

"AS STATED"
Sec. 4. "THE ACT"
Sec. 4.(b)(1) *Nothing in this Act shall apply to working conditions of employees with respect to which other Federal agencies, and State agencies acting under Section 274 of the Atomic Energy Act of 1954, as amended (42 U.S.C. 2021), exercise statutory authority to prescribe or enforce standards or regulations affecting occupational safety or health.*

DO THE OLD SAFETY AND HEALTH STANDARDS STILL APPLY?

No! Existing standards are superseded on the effective date of corresponding OSHA standards.

"AS STATED"
Sec. 4. "THE ACT"
Sec. 4.(b)(2) *The safety and health standards promulgated under the Act of June 30, 1936, commonly known as the Walsh–Healey Act (41 U.S.C. 35 et seq.), the Service Contract Act of 1965 (41 U.S.C. 351 et seq.), Public Law 91-54, Act of August 9, 1969 (40 U.S.C. 333), Public Law 85-742, Act of August 23, 1958 (33 U.S.C. 941), and the National Foundation on Arts and Humanities Act (20 U.S.C. 951 et seq.) are superseded on the effective date of corresponding standards, promulgated under this Act, which are determined by the Secretary to be more effective. Standards issued under the laws listed in this paragraph and in effect on or after the effective date of this Act shall be deemed to be occupational safety and health standards issued under this Act, as well as under such other Acts.*

HOW CAN ONE BE SURE THAT HE IS NOT VIOLATING ANOTHER LAW WHILE COMPLYING WITH OSHA?

It is the duty of the secretary to report to Congress to achieve coordination between this act and other federal laws.

> "AS STATED"
> Sec. 4. "THE ACT"
> Sec. 4.(b)(3) *The Secretary shall, within three years after the effective date of this Act, report to the Congress his recommendations for legislation to avoid unnecessary duplication and to achieve coordination between this Act and other Federal laws.*

Specal provisions of the law apply when some other federal law regulates the safety and health of employees.

The Labor Department: In some instances, your employees may have been covered by safety and health standards set by the Labor Department under an earlier law. In such a case, the standards set under OSHA replace any earlier standards dealing with the same occupational hazards. You are subject to enforcement procedures under both, but the Labor Department has said it will only proceed under one or the other, not both. The laws whose standards are replaced are the following:

1 / Walsh–Healey Public Contracts Act.
2 / Service Contract Act.
3 / Contract Work Hours and Safety Standards Act.
4 / Longshoremen's and Harbor Workers' Compensation Act.
5 / National Foundation on Arts and Humanities Act.

Other federal agencies: In other instances, some of your employees may be covered by occupational safety or health regulations set by a federal agency other than the Labor Department. As far as the working conditions of those employees are concerned, you should comply with the regulations of the other federal agency instead of the safety and health regulations under OSHA. The same exemption applies if a state agency exercises jurisdiction under the Atomic Energy Act of 1954.

OSHA DOES NOT CHANGE THE WORKMEN'S COMPENSATION LAW

OSHA does not in any manner affect the Workmen's Compensation Law.

> "AS STATED"
> Sec. 4. "THE ACT"
> Sec. 4.(b)(4) *Nothing in this Act shall be construed to supersede or in any manner affect any workmen's compensation law or to enlarge or diminish or affect in any other manner the common law or statutory rights, duties, or liabilities of employers and employees under any law with respect to injuries, diseases, or death of employees arising out of, or in the course of, employment.*

WHAT MUST BE DONE? THE LAW IS CLEAR

Every employer is required to furnish to all employees employment and places of employment free from recognized hazards causing or likely to cause death or serious physical harm; the employer has the specific duty of complying with safety and health standards promulgated under the act and all applicable rules, regulations, and orders issued pursuant to the act. (See Chapter 3.)

All employees are required to comply with all safety and health standards, and all rules, regulations, and orders issued pursuant to the act that are applicable to their actions and conduct. (See Chapter 3.)

HOW CAN THE RESPONSIBILITIES BE MET?

Establish and Enforce an Accident Prevention Safety Program

To effectively fulfill responsibilities and carry out the full meaning and intent of the act, an effective accident prevention safety program must be established and enforced [see the Act, Sec. 2.(b)(1)] covering every employee and workplace in the nation. Chapter 15 is a basic and practical accident prevention safety program.

DEFINITIONS

Chapter 2 covers definitions of terms used in the act.

RESPONSIBILITIES AND DUTIES OF EMPLOYERS AND EMPLOYEES

Chapter 3 covers the general responsibilities and duties of employers and employees.

INSPECTIONS AND INVESTIGATIONS

Inspections and investigations are covered in Chapter 4.

CITATIONS AND PENALTIES

The act provides for the issuing of citations and the levying of fines and/or imprisonment for anyone for certain violations. See Chapter 5.

ENFORCEMENT AND APPEALS

The act provides for enforcement and appeals of standards, citations, etc. See Chapter 6.

VARIANCES

The act provides for consideration and allowances of variances under certain conditions. See Chapter 7.

DISCRIMINATION

The act specifically guards against discrimination. See Chapter 8.

ACCIDENT REPORTING REQUIREMENTS

The act requires certain accidents to be reported immediately. The requirements for accident reporting are covered in Chapter 9.

RECORDKEEPING

Recordkeeping requirements are covered in Chapter 10.

INFORMATION AND POSTING REQUIREMENTS

The furnishing of information to employees and mandatory posting requirements are covered in Chapter 11.

SAFETY AND HEALTH STANDARDS

The act provides for the development of and compliance with safety and health standards to achieve improved safe and healthful working conditions. The standards are quite extensive and are undergoing constant updating and revision. To assist employers and employees in identifying those standards which affect their place of employment, Chapter 12 includes a guide to safety and health standards.

FEDERAL AGENCY PROGRAMS

Provisions for federal government agency programs are covered in Chapter 13.

ADDITIONAL PROVISIONS

Helpful inclusions and references to additional provisions in the act are covered in Chapter 14.

BRINGING YOUR WORKPLACE INTO COMPLIANCE

Guidelines for bringing your workplace into compliance are included in Chapter 16.

CHANGES

Attention must be given from time to time to changes that occur in the law and/or requirements. Consult the nearest OSHA office (see the Directory at the back of this manual) for such changes.

2

DEFINITIONS OF TERMS

DEFINITIONS OF TERMS

Definitions of the terms used herein are as follows:

Sec. 3. "The Act"

"for the purposes of this Act"

Secretary

Sec. 3.(1) *The term "Secretary" means the Secretary of Labor.*

Commission

Sec. 3.(2) *The term "Commission" means the Occupational Safety and Health Review Commission established under this Act.*

Commerce

Sec. 3.(3) *The term "commerce" means trade, traffic, commerce, transportation, or communication among the several States, or between a State and any place outside thereof, or within the District of Columbia, or a possession of the United States (other than the Trust Territory of the Pacific Islands), or between points in the same State but through a point outside thereof.*

Person

Sec. 3.(4) *The term "person" means one or more individuals, partnerships, associations, corporations, business trusts, legal representatives, or any group of persons.*

Employer

Sec. 3.(5) *The term "employer" means a person engaged in a business affecting commerce who has employees, but does not include the United States or any State or political subdivision of a State.*

Employee

Sec. 3.(6) *The term "employee" means an employee of an employer who is employed in a business of his employer which affects commerce.*

State

Sec. 3.(7) *The term "State" includes a State of the United States, the District of Columbia, Puerto Rico, the Virgin Islands, American Samoa, Guam, and the Trust Territory of the Pacific Islands.*

Occupational Safety and Health Standard

Sec. 3.(8) *The term "occupational safety and health standard" means a standard which requires conditions, or the adoption or use of one or more practices, means, methods, operations, or processes, reasonably necessary or appropriate to provide safe or healthful employment and places of employment.*

National Consensus Standard

Sec. 3.(9) *The term "national consensus standard" means any occupational safety and health standard or modification thereof which*

(1) has been adopted and promulgated by a nationally recognized standards-producing organization under procedures whereby it can be determined by the Secretary

that persons interested and affected by the scope or provisions of the standard have reached substantial agreement on its adoption,

(2) was formulated in a manner which afforded an opportunity for diverse views to be considered and

(3) has been designated as such a standard by the Secretary, after consultation with other appropriate Federal agencies.

Establisehd Federal Standard

Sec. 3.(10) *The term "established Federal standard" means any operative occupational safety and health standard established by any agency of the United States and presently in effect, or contained in any Act of Congress in force on the date of enactment of this Act.*

Committee

Sec. 3.(11) *The term "Committee" means the National Advisory Committee on Occupational Safety and Health established under this Act.*

Director

Sec. 3.(12) *The term "Director" means the Director of the National Institute for Occupational Safety and Health.*

Institute

Sec. 3.(13) *The term "Institute" means the National Institute for Occupational Safety and Health established under this Act.*

Workmen's Compensation Commission

Sec. 3.(14) *The term "Workmen's Compensation Commission" means the National Commission on State Workmen's Compensation Laws established under this Act.*

Serious Violation

Sec. 17(k) *For purposes of this section, a serious violation shall be deemed to exist in a place of employment if there is a substantial probability that death or serious physical harm could result from a condition which exists, or from one or more practices, means, methods, operations, or processes which have been adopted or are in use, in such place of employment unless the employer did not, and could not with the exercise of reasonable diligence, know of the presence of the violation.*

3

RESPONSIBILITIES AND DUTIES OF EMPLOYERS AND EMPLOYEES

RESPONSIBILITIES AND DUTIES OF EMPLOYERS AND EMPLOYEES

"EMPLOYER'S" RESPONSIBILITIES AND DUTIES

WHO IS CONSIDERED AN EMPLOYER?

You are considered an employer if you have employees unless you are the U.S. government or any state or political subdivision of a state.

> • **NOTE** • For governmental agency safety programs and responsibilities, see Chapter 13.

"AS STATED"
"Definition": *The term "employer" means any person engaged in a business affecting commerce who has employees, but does not include the United States or any State or political subdivision of a State.*

"Definition": *The term "person" means one or more individuals, partnerships, associations, corporations, business trusts, legal representative, or any organized group of persons.*

EACH EMPLOYER SHALL FURNISH SAFE AND HEALTHFUL WORKPLACES

"AS STATED"
Sec. 5. "THE ACT"
Sec. 5.(a)(1) *Each employer shall furnish to each of his employees employment and a place of employment which are free from recognized hazards that are causing or are likely to cause death or serious physical harm to his employees.*

This general duty clause has been described as a "catch-all provision." It imposes a duty without clear-cut limits. Just exactly what is required will depend on how OSHA and the courts interpret two key phrases: "recognized hazards" and "causing or likely to cause death or serious physical harm."

"Recognized hazards"

Congressman Dominick V. Daniels of New Jersey, who introduced this phrase, said that a "recognized hazard" is a condition generally known to be hazardous—it does not matter what you personally may recognize as a hazard or what a particular inspector may recognize as a hazard. Whether or not a condition is hazardous depends on "the standard of knowledge in the industry." OSHA Review Commission Chairman Robert D. Moran said that he is likely to be looking for evidence which will show, as a minimum, that the hazard (1) can be readily detected with the use of only the basic human senses, and (2) would be recognized by all "reasonably prudent" people as a hazard likely to cause death or serious physical harm.

The *OSHA Field Operations Manual* says that a hazard is "recognized if it is a condition (1) of common knowledge or general recognition in the particular industry in which it occurs, and (2) detectable (a) by means of the senses (sight, smell, touch, and hearing), or (b) is of such wide, general recognition as a hazard in the industry that, even if it is not detectable by means of the senses, there are generally known and accepted tests for its existence that the employer should know about. For example, an excessive concentration of a toxic substance in the air would be a recognized hazard if it could be detected through the use of measuring devices. However, if a specific standard covered the hazard, the general duty clause would not be applicable.

"Causing or likely to cause death or serious physical harm"

There is no question that serious physical harm means at least something more than a stubbed toe or a nicked finger. But there are some injuries that cannot be clearly defined as "serious" or "not serious."

Serious versus nonserious violations: The *OSHA Field Operations Manual* classifies violations as "serious" or "nonserious" depending on whether there is a substantial probability that death or serious physical harm could result.

Another problem with the phrase is "likely to cause." In a case involving the phrase "substantial probability" in a general duty violation, an OSHA Review Commission judge said it was enough to show that death or serious physical harm could result from a hazardous condition. The chance of an accident was enough to require the employer to take special precautions. The judge rejected the employer's argument for a 50 per cent probability test, pointing out that this was not required by the law and that the 50 per cent figure meant nothing without specifying the length of time the hazard existed. Another Review Commission judge said it was enough that a "potential" for serious injury existed, even if no injury actually resulted.

Violations of the general duty clause subject employers to the same penalties and enforcement procedure as violations of specific standards with one exception: citations based on the general duty clause will be limited to alleged serious violations (including willful and/or repeated violations that would otherwise qualify as serious violations).

Each Employer Shall Comply with Rules and Regulations

"AS STATED"
Sec. 5. "THE ACT"
Sec. 5.(a)(2) *Each employer shall comply with occupational safety and health standards promulgated under this Act.*

Each Employer Shall Provide Informaton to Employees and Post Required Notices

See Chapter 11.

Each Employer Shall Report Accidents Promptly

See Chapter 9.

Each Employer (with Eight or More Employees) Must Maintain Required Records

See Chapter 10.

Each Employer Must Enact and Maintain an Effective Accident Prevention Safety Program

To effectively fulfill responsibilities in ensuring compliance and to carry out the full meaning and intent of the law, an effective accident prevention safety program must be established [see Sec. 2.(b) (1) of the act, cross indexed at the front of this manual] by every employer covering every employee and workplace of the employer. Chapter 15 is a comprehensive practical accident prevention safety program.

Are Employers Responsible for Accidents Caused by Employees?

An employer normally will not be held responsible for an accident caused by an employee's disregard of safety rules, if there was no reasonable way to foresee the employee's action. If work rules for safety are well established and if the employees know about them and disregard them, the employer probably will not be held in violation—if he could not reasonably predict that the employees would not cooperate. But if the employer knows that his employees are not following safety rules and procedures, he will get a citation.

A company is responsible for the acts of its supervisors, even if they are violating company rules. If a company lets a supervisor break work rules or wink at noncompliance with safety rules, the company will be held responsible.

Contractors and Subcontractors

Under the general duty clause, all employers have a duty to provide their employees with a safe place to work. This applies to contractors and subcontractors, whether working on their own property or on that of another. The Review Commission has held that when a company sends its employees on another's property, that property becomes the company's workplace.

Cannot contract duty away

An employer cannot contract out his statutory duty to someone else. For example, an employer is not relieved from responsibility under the act just because other contractors and subcontractors may have caused an accident and may even be legally liable for damages to injured persons. A contractor working on the premises of another cannot avoid liability by an agreement with the owner of the premises to assume responsibility for safety and "hold harmless" the contractor of any liability.

Whose employees were exposed to a hazard? That is the question asked by Review Commission judges and the review commissioners when contractor–subcontractor disputes come before them. In general, the employer who controls the job situations is the employer who is responsible for safety. In the case of an "independent contractor" working on a daily-rate basis, if the employer directs the contractor in the performance of his job, he will be considered the employer and will be responsible for the contractor's actions.

General contractors

A general contractor is not automatically liable for all conditions on a job. If a hazard is created by a subcontractor and the general contractor has no employees affected by the hazard, he is not in violation.

On the other hand, subcontractors are not responsible for hazardous conditions caused by general contractors if the subcontractor's employees are not exposed to the hazard.

Two or more employers may be held responsible if all have employees working under unsafe conditions.

EMPLOYEE'S RESPONSIBILITIES AND DUTIES

Who Is Considered an Employee?

You are considered an employee if you are employed by an employer other than the U.S. government or any state or political subdivision of a state.

> • **NOTE** • For employees covered under governmental agency safety programs and responsibilities, see Chapter 13.

"AS STATED"
"Definition": *The term "employee" means an employee of an employer who is employed in a business of his employer which affects commerce.*

Each Employee Shall Comply with Rules and Regulations

"AS STATED"
Sec. 5. "THE ACT"
Sec. 5.(b) *Each employee shall comply with occupational safety and health standards and all rules, regulations, and orders issued pursuant to this Act which are applicable to his own actions and conduct.*

The act provides that employees "shall comply with occupational safety and health standards and all rules, regulations, and orders" that are applicable to them. However, the act sets up no means of enforcing this duty and no penalties for not complying.

Failure of employees to comply does not affect the obligations of employers under the act. But to what extent can employers be held responsible for employees' failure to observe safety regulations? There's no hard and fast rule; generally, an employer will be found in violation only if he does something he should not do or fails to do something he should. However, if he knows or should know that employees are not following safety rules, he has a duty to see that they are enforced.

Use of personal protective equipment: Where personal protective equipment is required, it is not enough for an employer to just provide it; he has to make sure that it is used. At a worksite where noise levels were excessive unless earplugs were used, an employer provided the equipment and instructed employees in hearing safety through safety meetings, a safety manual, signs, etc. Nevertheless, a Review Commission judge found a violation; although employees had an obligation to cooperate, the employer should have required the use of earplugs.

But you may not have to pay for it: The act says you have to provide personal protective equipment and make sure employees use it, but the Review Commission has ruled that the act does not require the employer to pay for it. The commission said it is irrelevant who pays for the equipment; the question of cost allocation is a matter to be resolved between the employer and his employees.

Each Employee Must Comply with Employer's Accident Prevention Safety Program

An effective accident prevention safety program must be established and maintained covering each place of employment. It is each employee's responsibility and duty to participate in and comply with all accident prevention safety programs affecting his workplace.

CITATIONS AND PENALTIES

There are provisions in the law for citations and penalties for violations of the law. (See Chapter 5.)

RIGHTS OF EMPLOYERS AND EMPLOYEES

The law provides for certain and specific rights of the employer and employee, for example, the right to request a variance, to request an inspection, to file a complaint, and to appeal or contest a citation. To ascertain your rights under the law, refer to the various subject chapter headings of this manual, i.e., variances, inspections and investigations, citations and penalties, discrimination, etc.

4

INSPECTIONS AND INVESTIGATIONS

INSPECTIONS AND INVESTIGATIONS

The law provides for the inspection and investigation of all workplaces as follows:

RIGHT TO INSPECT AND INVESTIGATE

BARROWS CASE

The OSHA inspector has the right to enter your workplace without delay and to inspect the conditions therein relative to compliance with OSHA requirements.

The OSHA inspector may question in private any employer, owner, operator, agent, or employee.

"AS STATED"
Sec. 8. "THE ACT"
Sec. 8.(a) *In order to carry out the purposes of this Act, the Secretary, upon presenting appropriate credentials to the owner, operator, or agent in charge, is authorized*

(1) to enter without delay and at reasonable times any factory, plant, establishment, construction site, or other area, workplace, or environment where work is performed by an employee of an employer; and

(2) inspect and investigate during regular working hours and at other reasonable times, and within reasonable limits and in a reasonable manner, any such place of employment and all pertinent conditions, structures, machines, apparatus, devices, equipment, and materials therein, and to question privately any such employer, owner, operator, agent, or employee.

WHEN WILL AN OSHA INSPECTOR CALL?

A system of priorities has been established for inspections as follows:

1 / Catastrophes and other fatal accidents.
2 / Valid employee complaints.
3 / Special-emphasis programs: target industries and target health hazards.
4 / Random selection from all types and sizes of workplaces in all sections of the country.

• **NOTE** • You should at all times be ready for a compliance inspection at each of your workplaces.

YOU WILL NOT BE NOTIFIED OF AN INSPECTION OF YOUR WORKPLACE

You will not know you are to be inspected until the compliance officer (inspector) arrives (except under special circumstances). Advance notice of inspection is prohibited by law (except under special circumstances) and carries a penalty of up to $1,000 and/or imprisonment of up to six months. (See Chapter 5.)

OSHA INSPECTORS

Inspections will be made by OSHA compliance safety and health officers and industrial hygienists.

HOW IS AN OSHA INSPECTION CONDUCTED?

OSHA inspections are conducted during regular working hours of the establishment to be inspected except in special circumstances, where advance notice of inspections would serve to make the inspection more effective. When the compliance officer arrives at the establishment to be inspected, he will present himself, display his credentials, and ask to meet the appropriate employer representative.

The compliance officer will then inform the employer of the reason for his visit and outline in general terms the scope of the inspection to be made. He will inform the employer of the safety and health records that he may wish to review. The compliance officer will inform the employer of employee interviews that he will require during the inspection. He will explain in general terms the walk-around inspection and the closing conference.

The compliance officer will inform the employer of applicable laws and safety and health standards under which the inspection is to be conducted. He will give to the employer a copy of an employee complaint if one is involved. The employee's name will be withheld if the employee so requests.

The compliance officer will require the employer or a representative designated by the employer to be present for the walk-around inspection. He will also request that an employee representative be designated for the walk-around inspection. The employee representative shall be selected *by the employees* (not by the employer) from a union or unions or employment group in the workplace. If there are no employee groups, the compliance officer will discuss conditions with individual employees during the walk-around inspection.

THE OSHA INSPECTION

The compliance officer, the employer (or designated employer representative), and the employee representative will tour through the establishment with each work area being inspected for compliance with OSHA standards. No one may obstruct the inspection process in any way.

The compliance officer will take appropriate notes of conditions and discuss them with the representatives. He may also take photographs of particular situations to record apparent violations or conditions, and he may use other appropriate investigative techniques.

The compliance officer must take special care to protect the privacy of trade secrets or security matters.

During the walk-around inspection, obvious and apparent violations may be found that can be corrected immediately. Such violations could include blocked aisles, unsafe floor surfaces, hazardous projections, unsanitary conditions, etc. The employer representative may, and usually does, direct that they be corrected at once. Such corrections are recorded to help in judging the employer's good faith in compliance. Even though corrected, the apparent violation may be the basis for a citation and/or proposed penalty.

During the walk-around inspection any employee may bring to the attention of the compliance officer any condition he believes to be a violation.

The compliance officer will inspect the required employer-maintained OSHA records (Forms 100, 101, and 102) of deaths, injuries, and illnesses, and determine that the required posting has been accomplished. The compliance officer will also check records of employee exposure to toxic substances and harmful physical agents.

Following the walk-around inspection, the compliance officer will discuss with the employer what he has seen and review probable violations. Also, he will discuss the time the employer believes he will need to abate (correct) the hazards found.

Following the inspection, the compliance officer will return to his office and file his report. Citations (if any) to be issued will be determined and penalties (if any) will be proposed. Copies of citations and proposed penalties will be sent by certified mail to the employer with a copy to the complainant, if there was one. The compliance officer may not, on his own, impose or propose a penalty "on the spot" at an inspection.

OSHA citations and proposed penalties are similar to traffic violations. If contested, they are subject to final action by a separate authority. (See Chapter 6.)

OSHA inspection procedure may vary somewhat according to the inspection being made. However, the preceding procedures will in general be followed.

YOUR ATTITUDE AND COOPERATION DURING AN INSPECTION

Your attitude and cooperation during an OSHA inspection toward compliance with this mandatory law, which everyone must comply with, will definitely have an affect on the outcome of the inspection. OSHA compliance officers are professionals who fully recognize that a person's attitude and cooperation toward attaining maximum safety are possibly the most important ingredients.

When the compliance officer arrives to inspect your establishment:

DON'T keep him waiting for any length of time.
DON'T try to prohibit him from entry.
DON'T ask him to come back.
DON'T expound your philosophy on the OSHA law.
DON'T try to harass or obstruct the inspection in any way.

DO cooperate.
DO provide whatever you can to make the inspection move as quickly and as easily as possible.
DO show the compliance officer your accident prevention safety program.
DO show a sincere interest to discover, eliminate, and prevent future safety and health hazards.

Remember: Your attitude and cooperation toward compliance with safety and health standards is important.

SUBPOENA POWER UNDER THE LAW

You can be required to testify and/or produce evidence under oath. If necessary, the courts have the power to subpoena your testimony in court.

"AS STATED"
Sec. 8. "THE ACT"
Sec. 8.(b) *In making his inspections and investigations under this Act, the Secretary may require the attendance and testimony of witnesses and the production of evidence under oath. Witnesses shall be paid the same fees and mileage that are paid witnesses in the courts of the United States. In case of a contumacy, failure, or refusal of any person to obey such an order, any district court of the United States or the United States courts of any territory or possession, within the jurisdiction of which such person is found, or resides or transacts business, upon the application by the Secretary, shall have jurisdiction to issue to such person an order requiring such person to appear to produce evidence if, as, and when so ordered, and to give testimony relating to the matter under investigation or in question, and any failure to obey such order of the court may be punished by said court as a contempt thereof.*

YOUR RIGHT TO ACCOMPANY THE OSHA INSPECTOR DURING AN INSPECTION

You or your designated representative are given the right and opportunity to accompany the OSHA inspector on his inspection.

"AS STATED"
Sec. 8. "THE ACT"
Sec. 8.(e) *Subject to regulations issued by the Secretary, a representative of the employer and a representative authorized by his employees shall be given an opportunity to accompany the Secretary or his authorized representative during the physical inspection of any workplace under subsection (a) for the purpose of aiding such inspection. Where there is no authorized employee representative, the Secretary or his authorized representative shall consult with a reasonable number of employees concerning matters of health and safety in the workplace.*

WALK-AROUND PAY

The Labor Department has stated that you do not have to pay your employees for the time they spend participating in the walk-around if you do not normally pay employees for similar activities. However, if you usually pay employees for the time they spend conducting your own safety inspections, you may be required to pay for walk-around time spent with an OSHA inspector. The Labor Department's position may change, though, since it is being appealed to the courts.

EMPLOYEE'S RIGHT TO REQUEST INSPECTION

Employees or their authorized representative can request an inspection if they feel that a violation of a safety or health standard exists that threatens physical harm, or that an imminent danger exists, by sending a written request to the secretary of labor. The request for inspection must explain in detail the nature of the violation or the hazard, and must be signed by the employee or the representative. The employer has a right to receive a copy of the complaint at the time of the inspection, but the complainant's name will be blanked out if requested. If the secretary determines that an inspection is not warranted, he must notify the employees or representative in writing.

> "AS STATED"
> Sec. 8. "THE ACT"
> Sec. 8.(f)(1) *Any employees or representative of employees who believe that a violation of a safety or health standard exists that threatens physical harm, or that an imminent danger exists, may request an inspection by giving notice to the Secretary or his authorized representative of such violation or danger. Any such notice shall be reduced to writing, shall set forth with reasonable particularity the grounds for the notice, and shall be signed by the em-*

ployees or representative of employees, and a copy shall be provided the employer or his agent no later than at the time of inspection, except that, upon the request of the person giving such notice, his name and the names of individual employees referred to therein shall not appear in such copy or on any record published, released, or made available pursuant to subsection (g) of this section. If upon receipt of such notification the Secretary determines there are reasonable grounds to believe that such violation or danger exists, he shall make a special inspection in accordance with the provisions of this section as soon as practicable, to determine if such violation or danger exists. If the Secretary determines there are no reasonable grounds to believe that a violation or danger exists he shall notify the employees or representative of the employees in writing of such determination.

Sec. 8.(f)(2) *Prior to or during any inspection of a workplace, any employees or representative of employees employed in such workplace may notify the Secretary or any representative of the Secretary responsible for conducting the inspection, in writing, of any violation of this Act which they have reason to believe exists in such workplace. The Secretary shall, by regulation, establish procedures for informal review of any refusal by a representative of the Secretary to issue a citation with respect to any such alleged violation and shall furnish the employees or representative of employees requesting such review a written statement of the reasons for the Secretary's final disposition of the case.*

PUBLICATION OF REPORTS

OSHA may publish all reports and information obtained during an inspection, except trade secrets (see the Act, Sec. 15).

"AS STATED"
Sec. 8. "THE ACT"
Sec. 8.(g)(1) *The Secretary and Secretary of Health, Education, and Welfare are authorized to compile, analyze,*

and publish, either in summary or detailed form, all reports or information obtained under this section.

PROVISION FOR PRESCRIBING RULES AND REGULATIONS FOR INSPECTIONS

"AS STATED"
Sec. 8. "THE ACT"
Sec. 8.(g)(2) *The Secretary and the Secretary of Health, Education, and Welfare shall each prescribe such rules and regulations as he may deem necessary to carry out their responsibilities under this Act, including rules and regulations dealing with the inspection of an employer's establishment.*

PROTECTION OF TRADE SECRETS

All trade secrets must be kept confidential.

"AS STATED"
Sec. 15. "THE ACT"
Sec. 15. *All information reported to or otherwise obtained by the Secretary or his representative in connection with any inspection or proceeding under this Act which contains or which might reveal a trade secret referred to in Section 1905 of Title 18 of the United States Code shall be considered confidential for the purpose of that section, except that such information may be disclosed to other officers or employees concerned with carrying out this Act or when relevant in any proceeding under this Act. In any such proceeding the Secretary, the Commission, or the court shall issue such orders as may be appropriate to protect the confidentiality of trade secrets.*

ENFORCEMENT AND APPEALS

For enforcement and right of appeal, see Chapter 6.

5

CITATIONS AND PENALTIES

CITATIONS AND PENALTIES

There are provisions in the law for citations and penalties for violations as follows:

YOU WILL RECEIVE A CITATION FOR VIOLATION

The OSHA inspector must issue a citation in writing for each violation detected. Each citation must state the nature of the violation and give a reference to the standard, rule, or regulation violated, as well as a reasonable time for correcting the violation. You may receive a notice (in lieu of a citation) for minor violations that do not immediately or directly affect safety or health.

> "AS STATED"
> Sec. 9. "THE ACT"
> Sec. 9.(a) *If, upon inspection or investigation, the Secretary or his authorized representative believes that an employer has violated a requirement of Section 5 of this Act, of any standard, rule, or order promulgated pursuant to Section 6 of this Act, or of any regulations prescribed pursuant to this Act, he shall with reasonable promptness issue a citation to the employer. Each citation shall be in writing and shall describe with particularity the nature of the violation, including a reference to the provision of the Act, standard, rule, regulation, or order alleged to have been violated. In addition, the citation shall fix a reasonable time for the abatement of the violation. The Secretary may prescribe procedures for the issuance of a notice in lieu of a citation with respect to de minimis violations which have no direct or immediate relationship to safety or health.*

DE MINIMIS VIOLATIONS

No citation will be issued if a violation of a standard has no immediate or direct relationship to safety or health. In these situations, a notice of de minimis violation will be issued with no proposed penalty.

YOU MUST POST A COPY OF EACH CITATION

The act provides that every citation must be prominently posted at or near each place a violation has allegedly occurred. Regulations require it to be posted for at least three days or until the violation is corrected, whichever is longer.

Contested citations: The employer has to post the citation, even if he is contesting it. However, he can also post a notice stating that he is contesting the citation.

"AS STATED"
Sec. 9. "THE ACT"
Sec. 9.(b) *Each citation issued under this section, or a copy or copies thereof, shall be prominently posted, as prescribed in regulations issued by the Secretary, at or near each place a violation referred to in the citation occurred.*

CITATIONS MUST BE ISSUED PROMPTLY

"AS STATED"
Sec. 9. "THE ACT"
Sec. 9.(c) *No citation may be issued under this section after the expiration of six months following the occurrence of any violation.*

PENALTIES

Repeated violation	$10,000 each maximum
Serious violation	$1,000 each maximum
Nonserious violation	$1,000 each maximum
Failure to correct	$1,000 per day maximum
Willful violation	$10,000 or 6 months and $20,000 or 1 year, or both
Giving advance notice of inspection	$1,000 or 6 months or both
Falsification	$10,000 or 6 months or both
Killing law enforcement personnel	Up to life imprisonment
Violation of posting requirements	$1,000 each maximum

PENALTY PROVISIONS

Repeated Violation

$10,000 each maximum. You can be fined up to $10,000 for willful or repeated violations.

> "AS STATED"
> Sec. 17. "THE ACT"
> Sec. 17.(a) *Any employer who willfully or repeatedly violates the requirements of Section 5 of this Act, any standard, rule, or order promulgated pursuant to Section 6 of this Act, or regulations prescribed pursuant to this Act, may be assessed a civil penalty of not more than $10,000 for each violation.*

A citation for a repeated violation will be issued if an employer violated the same standard, rule, order, or the general duty clause if he has already been previously cited for the same violation. Repeated violations differ from willful violations in that they may result from inadvertent, accidental, or ordinarily negligent acts. Repeated violations differ from failure to abate in that repeated violations exist if the employer has already abated the earlier violation, and, upon later inspection, is found to have violated the same standard again. A notice of failure to abate would be appropriate if the employer has been cited and fails to abate the hazard within the specified abatement period.

Serious Violation

$1,000 each maximum. You can be fined up to $1,000 for each serious violation (see definition of "serious violation" below).

> "AS STATED"
> Sec. 17. "THE ACT"
> Sec. 17.(b) *Any employer who has received a citation for a serious violation of the requirements of Section 5 of this Act, of any standard, rule, or order promulgated pursuant to Section 6 of this Act, or of any regulations prescribed pursuant to this Act, shall be assessed a civil penalty of up to $1,000 for each such violation.*
>
> Sec. 17.(k) *For purposes of this section, a serious violation shall be deemed to exist in a place of employment if there is a substantial probability that death or serious physical harm could result from a condition which exists, or from one or more practices, means, methods, operations, or processes which have been adopted or are in use, in such place of employment unless the employer did not, and could not with the exercise of reasonable diligence, know of the presence of the violation.*

A serious violation exists if there is "substantial probability" that the consequences of an accident resulting from the violation will be death or serious physical harm, unless the employer did not and could not with the exercise of "reasonable diligence" know the hazard was present. In deciding whether a violation is serious, both compliance officer and area director must determine the following:

1 / What is serious physical harm? Serious physical harm is harm that would cause permanent or prolonged impairment of the body or that, although not impairing the body on a prolonged basis, could cause such temporary disablement that would warrant hospitalization. Serious physical harm also includes less obvious harm that could inhibit an internal bodily system in the performance of its normal function to such a degree as to shorten life or cause reduction in physical or mental efficiency.

2 / Likelihood of injury: The inspector must consider the most likely results of the hazards when assessing serious physical harm, and the assessment must be independent of any consideration of what effect abatement or medical treatment would have on the injury.

3 / Employer's knowledge: A violation will not be considered "serious" if the employer "did not, and could not with the exercise of reasonable diligence, know of the presence of the violation." The knowledge test is met if the inspector is satisfied that the employer actually knew of the condition that constituted the violation. In many cases, it will be difficult to determine employer knowledge. If that is the case, the "reasonable diligence" test must be applied. Assuming that the employer is safety conscious and possesses the technical expertise normally expected of an employer engaged in that particular business, the reasonable diligence requirement will be met if the employer should have known of the violation.

Compound violations: A citation for a serious violation may be issued for a group of individual violations, which, taken by themselves, would not be classed as serious, but considered together would present a serious hazard.

Nonserious Violation

$1,000 each maximum. You can be fined up to $1,000 for each nonserious violation.

"AS STATED"
Sec. 17. "THE ACT"
Sec. 17.(c) *Any employer who has received a citation for a violation of the requirements of Section 5 of this Act, of any standard, rule, or order promulgated pursuant to Section 6 of this Act, or of regulations prescribed pursuant to this Act, and such violation is specifically determined not to be of a serious nature, may be assessed a civil penalty of up to $1,000 for each such violation.*

Citations for nonserious violations will be issued in situations where an accident or occupational illness resulting from a violation of a standard would probably not cause death or serious physical harm, but would have a direct or immediate effect on the safety or health of employees. An example of a nonserious violation is the lack of guardrails at a height from which a fall would more probably result in only a mild sprain or cuts and abrasions, i.e., something less than serious physical harm.

Failure to Correct

$1,000 per day maximum. You can be fined $1,000 per day for failing to correct a violation.

"AS STATED"
Sec. 17. "THE ACT"
Sec. 17.(d) *Any employer who fails to correct a violation for which a citation has been issued under Section 9(a) within the period permitted for its correction (which period shall not begin to run until the date of the final order of the Commission in the case of any review proceeding under Section 10 initiated by the employer in good faith and not solely for delay or avoidance of penalties), may be assessed a civil penalty of not more than $1,000 for each day during which such failure or violation continues.*

Willful Violation

$10,000 or 6 months, $20,000 or 1 year. For willful violations you can be fined $10,000 and/or 6 months in prison for the first offense and $20,000 and/or 1 year in prison for repeated willful violations.

> "AS STATED"
> Sec. 17. "THE ACT"
> Sec. 17.(e) *Any employer who willfully violates any standard, rule, or order promulgated pursuant to Section 6 of this Act, or of any regulations prescribed pursuant to this Act, and that violation caused death to any employee, shall, upon conviction, be punished by a fine of not more than $10,000 or by imprisonment for not more than six months, or by both; except that if the conviction is for a violation committed after a first conviction of such person, punishment shall be by a fine of not more than $20,000 or by imprisonment for not more than one year, or by both.*

A citation for a willful violation will be issued if the evidence shows (1) that the employer committed an intentional and knowing violation and the employer knew that he violated the act, or (2) even though the employer was not consciously violating the act, he was aware that a hazardous condition existed and made no reasonable effort to eliminate the condition. Prior knowledge of the condition and lack of action to remove the exposure to employees is a valid reason to cite for a willful violation. Evil intent is not required; it is enough that the violation was deliberate, voluntary, or intentional.

Giving Advance Notice of Inspections

$1,000 or 6 months. Giving advance notice of inspections is prohibited and punishable by a fine of $1,000 and/or 6 months in prison.

"AS STATED"
Sec. 17. "THE ACT"
Sec. 17.(f) *Any person who gives advance notice of any inspection to be conducted under this Act, without authority from the Secretary or his designees, shall, upon conviction, be punished by a fine of not more than $1,000 or by imprisonment for not more than six months, or by both.*

Falsification

$10,000 or 6 months. You can be fined up to $10,000 and/or 6 months in prison for not telling the truth.

"AS STATED"
Sec. 17. "THE ACT"
Sec. 17.(g) *Whoever knowingly makes any false statement, representation, or certification in any application, record, report, plan, or other document filed or required to be maintained pursuant to this Act shall, upon conviction, be punished by a fine of not more than $10,000, or by imprisonment for not more than six months, or by both.*

Killing Enforcement Personnel

To life imprisonment. The lives of enforcement personnel are specifically protected.

"AS STATED"
Sec. 17. "THE ACT"
Sec. 17.(h)(1) *Section 1114 of Title 18, United States Code, is hereby amended by striking out "designated by the Secretary of Health, Education, and Welfare to conduct investigations, or inspections under the Federal Food, Drug, and Cosmetic Act" and inserting in lieu thereof "or of the Department of Labor assigned to perform investigative, inspection, or law enforcement functions."*

Sec. 17.(h)(2) *Notwithstanding the provisions of Sections*

1111 and 1114 of Title 18, United States Code, whoever, in violation of the provisions of Section 1114 of such title, kills a person while engaged in or on account of the performance of investigative, inspection, or law enforcement functions added to such Section 1114 by paragraph (1) of this subsection, and who would otherwise be subject to the penalty provisions of such Section 1111, shall be punished by imprisonment for any term of years or for life.

Violation of Posting Requirements

$1,000 each maximum. You can be fined up to $1,000 for violating the posting requirements. (See Chapter 11.)

"AS STATED"
Sec. 17. "THE ACT"
Sec. 17.(i) *Any employer who violates any of the posting requirements, as prescribed under the provisions of this Act, shall be assessed a civil penalty of up to $1,000 for each violation.*

COMMISSION'S AUTHORITY TO ASSESS PENALTIES

The commission has the authority to assess penalties. Due consideration must be given to the appropriateness of the penalty with respect to the size of the employer's business, the gravity of the violation, good faith, and past history of violations.

"AS STATED"
Sec. 17. "THE ACT"
Sec. 17.(j) *The Commission shall have authority to assess all civil penalties provided in this section, giving due consideration to the appropriateness of the penalty with respect to the size of the business of the employer being charged, the gravity of the violation, the good faith of the employer, and the history of previous violations.*

PAYMENT OF PENALTIES

Upon receiving a notice of penalty, you will be given instructions as to where to pay the penalty. Ultimately, the penalties go into the U.S. Treasury.

> "AS STATED"
> Sec. 17. "THE ACT"
> Sec. 17(l) *Civil penalties owed under this Act shall be paid to the Secretary for deposit into the Treasury of the United States and shall accrue to the United States and may be recovered in a civil action in the name of the United States brought in the United States district court for the district where the violation is alleged to have occurred or where the employer has its principal office.*

6

ENFORCEMENT AND RIGHT OF APPEAL

ENFORCEMENT AND RIGHT OF APPEAL

Provisions for enforcement of the law and right of appeal are as follows:

EMPLOYER'S RIGHT TO DISAGREE WITH CITATION AND/OR PROPOSED PENALTY

Any employer who disagrees with a citation and/or proposed penalty can:

1 / Immediately request an informal meeting with the area director to discuss the case. (See the Directory at the back of this manual for the nearest area director.)

2 / *Within fifteen working days* from receipt of a citation and/or proposed penalty, notify the area director *in writing* that you intend to contest the citation, abatement date, or proposed penalty to the Occupational Safety and Health Review Commission. The area director will then send the case to the Review Commission.

SIX STEPS TO CONTEST A CITATION

The chairman of the Occupational Safety and Health Review Commission has outlined the six steps you must follow to properly contest an occupational safety or health enforcement action. You must carefully observe all six points, as outlined in the Review Commission's rules of procedure:

1 / Notify the area director in charge of the local Labor Department office that initiated the action against you that you wish to contest. (You have fifteen working days from the time you receive the certified mail letter containing the department's

proposed penalty.) This notification by you is the "notice of contest."

2 / If any of the employees who work at the site of the alleged violation are unionized, a copy of your notice of contest must be served on the union.

3 / If any of the employees who work at the site of the alleged violation are not represented by a union, copies of your notice of contest must either (a) be posted in a place where employees will see it, consistent with the requirements for posting of citations, or (b) be personally served on any nonunion employee.

4 / There is no particular form prescribed for your notice of contest, but it must clearly identify the basis for its filing—the citation, notification of proposed penalty, or notification of failure to correct violation.

5 / Your notice of contest must also contain a listing of the names and addresses of those parties to the case who have been personally served with the notice, plus the address where it has been posted (when posting is required). The reason for this listing is so that the commission can determine whether you have fully complied with its notification rules.

6 / If any of the employees who work at the site of the alleged violation are not represented by a union and you do not have them personally served with a copy of the notice of contest, the copies of the notice of contest which are posted must specifically advise those employees that they may lose their right to participate in the case if they fail to properly identify themselves to the commission or hearing examiner before or at the beginning of the hearing.

REVIEW COMMISSION'S ACTION

The Occupational Safety and Health Review Commission, which handles all contested cases, has no connection with the U.S. Department of Labor. The Review Commission cannot act unless the notice of contest was filed in time. (The postmark will decide, if the notice is mailed.)

If the notice is filed in time, the Review Commission assigns the case to an administrative law judge. He can investigate and disallow the contest if he finds it legally invalid, or he can schedule a hearing, which will be held as close as possible to the employer's workplace. The Review Commission does not require that employers or employees be represented by attorneys.

The employer can accept or object to the judge's findings and, upon the request of any party, the judge's decision may be reviewed by the Review Commission itself, although it is not required to do so. Any member of the Review Commission can, on his own motion, order a review of any contested case.

Decisions of the commission may be appealed to the U.S. Circuit Court of Appeals for the circuit in which the case arose.

If the employer or employee does not contest within fifteen days of receipt of the citation, the OSHA action automatically becomes a final order of the Review Commission and is not subject to further appeal or review.

Employees have the right to contest to the Review Commission if they believe the period set by OSHA for abatement of a hazard is unreasonable. An employer may petition the Review Commission (through the area director) for modification of the abatement. Such petition shall be after the fifteen days but before the end of the abatement period.

Notification of Penalty and Employer's Right to Contest

You will be notified by certified mail of penalties assessed against you. You have fifteen working days from the receipt of the notice to file your intent to contest the citation or proposed penalty. For your own protection, put your notice of intent to contest in writing and mail by certified mail to the OSHA area director (see the Directory at the back of this manual) within the time limits specified.

"AS STATED"
Sec. 10. "THE ACT"
Sec. 10.(a) *If, after an inspection or investigation, the*

Secretary issues a citation under Section 9(a), he shall, within a reasonable time after the termination of such inspection or investigation, notify the employer by certified mail of the penalty, if any, proposed to be assessed under Section 17 and that the employer has fifteen working days within which to notify the Secretary that he wishes to contest the citation or proposed assessment of penalty. If, within fifteen working days from the receipt of the notice issued by the Secretary, the employer fails to notify the Secretary that he intends to contest the citation or proposed assessment of penalty, and no notice is filed by any employee or representative of employees under subsection (c) within such time, the citation and the assessment, as proposed, shall be deemed a final order of the Commission and not subject to review by any court or agency.

Notification of Failure to Correct

You must correct any violation within the time limits specified on your citation. If you fail to correct the violation within the specified time, you can receive another citation and penalty for failure to correct. You have fifteen working days from the receipt of this notice to file a notice of intent to contest the citation or proposed penalty. For your own protection, put your notice of intent to contest in writing and mail by certified mail to the OSHA area director (see the Directory at the back of this manual) within the time limits specified.

• **NOTE** • You may not file a contest solely for the purpose of delay or avoidance of penalties.

"AS STATED"
Sec. 10. "THE ACT"
Sec. 10.(b) *If the Secretary has reason to believe that an employer has failed to correct a violation for which a citation has been issued within the period permitted for its correction (which period shall not begin to run until the entry of a final order by the Commission in the case of any re-*

view proceedings under this section initiated by the employer in good faith and not solely for delay or avoidance of penalties), the Secretary shall notify the employer by certified mail of such failure and of the penalty proposed to be assessed under Section 17 by reason of such failure, and that the employer has fifteen working days within which to notify the Secretary that he wishes to contest the Secretary's notification or the proposed assessment of penalty. If, within fifteen working days from the receipt of notification issued by the Secretary, the employer fails to notify the Secretary that he intends to contest the notification or proposed assessment of penalty, the notification and assessment, as proposed, shall be deemed a final order of the Commission and not subject to review by any court or agency.

EMPLOYER'S NOTIFICATION TO CONTEST

If you file a notice to contest a citation, notification, or the time allotted to correct a violation, you will be given a hearing. As a result of the hearing, the Review Commission can affirm, modify, or vacate the citation or penalty, or may direct appropriate relief. If you cannot correct the violation in the prescribed time because of factors beyond your reasonable control, the commission can allot you additional time for correcting the violation.

"AS STATED"
Sec. 10. "THE ACT"
Sec. 10.(c) *If an employer notifies the Secretary that he intends to contest a citation issued under Section 9(a) or notification issued under subsection (a) or (b) of this section, or if, within fifteen working days of the issuance of a citation under Section 9(a), any employee or representative of employees files a notice with the Secretary alleging that the period of time fixed in the citation for the abatement of the violation is unreasonable, the Secretary shall immediately advise the Commission of such notification, and the Commission shall afford an opportunity for a hearing*

(in accordance with Section 554 of Title 5, United States Code, but without regard to subsection (a)(3) of such section). The Commission shall thereafter issue an order, based on findings of fact, affirming, modifying, or vacating the Secretary's citation or proposed penalty, or directing other appropriate relief, and such order shall become final thirty days after its issuance. Upon a showing by an employer of a good faith effort to comply with the abatement requirements of a citation, and that abatement has not been completed because of factors beyond his reasonable control, the Secretary, after an opportunity for a hearing as provided in this subsection, shall issue an order affirming or modifying the abatement requirements in such citation. The rules of procedure prescribed by the Commission shall provide affected employees or representatives of affected employees an opportunity to participate as parties to hearings under this subsection.

Judicial Review

Should you wish to appeal the decision of the Review Commission, you may do so with the U.S. Court of Appeals within sixty days following the decision of the commission. The court of appeals can grant temporary relief or restraining order. The court of appeals can affirm, modify, or set aside all or any part of the order of the commission.

"AS STATED"
Sec. 11. "THE ACT"
Sec. 11.(a) *Any person adversely affected or aggrieved by an order of the Commission issued under subsection (c) of Section 10 may obtain a review of such order in any United States court of appeals for the circuit in which the violation is alleged to have occurred or where the employer has its principal office, or in the Court of Appeals for the District of Columbia Circuit, by filing in such court within sixty days following the issuance of such order a written petition praying that the order be modified or set aside. A*

copy of such petition shall be forthwith transmitted by the clerk of the court to the Commission and to the other parties, and thereupon the Commission shall file in the court the record in the proceeding as provided in Section 2112 of Title 28, United States Code. Upon such filing, the court shall have jurisdiction of the proceeding and of the question determined therein, and shall have power to grant such temporary relief or restraining order as it deems just and proper, and to make and enter upon the pleadings, testimony, and proceedings set forth in such record a decree affirming, modifying, or setting aside in whole or in part, the order of the Commission and enforcing the same to the extent that such order is affirmed or modified. The commencement of proceedings under this subsection shall not, unless ordered by the court, operate as a stay of the order of the Commission. No objection that has not been urged before the Commission shall be considered by the court, unless the failure or neglect to urge such objection shall be excused because of extraordinary circumstances. The findings of the Commission with respect to questions of fact, if supported by substantial evidence on the record considered as a whole, shall be conclusive. If any party shall apply to the court for leave to adduce additional evidence and shall show to the satisfaction of the court that such additional evidence is material and that there were reasonable grounds for the failure to adduce such evidence in the hearing before the Commission, the court may order such additional evidence to be taken before the Commission and to be made a part of the record. The Commission may modify its findings as to the facts, or make new findings, by reason of additional evidence so taken and filed, and it shall file such modified or new findings, which findings with respect to questions of fact, if supported by substantial evidence on the record considered as a whole, shall be conclusive, and its recommedations, if any, for the modification or setting aside of its original order. Upon the filing of the record with it, the jurisdiction of the court shall be exclusive and its judgment and decree shall be final, except that the same shall be sub-

ject to review by the Supreme Court of the United States, as provided in Section 1254 of Title 28, United States Code. Petitions filed under this subsection shall be heard expeditiously.

Sec. 11.(b) *The Secretary may also obtain review or enforcement of any final order of the Commission by filing a petition for such relief in the United States court of appeals for the circuit in which the alleged violation occurred or in which the employer has its principal office, and the provisions of subsection (a) shall govern such proceedings to the extent applicable. If no petition for review, as provided in subsection (a), is filed within sixty days after service of the Commission's order, the Commission's findings of fact and order shall be conclusive in connection with any petition for enforcement which is filed by the Secretary after the expiration of such sixty-day period. In any such case, as well as in the case of a noncontested citation or notification by the Secretary which has become a final order of the Commission under subsection (a) or (b) of Section 10, the clerk of the court, unless otherwise ordered by the court, shall forthwith enter a decree enforcing the order and shall transmit a copy of such decree to the Secretary and the employer named in the petition. In any contempt proceeding brought to enforce a decree of a court of appeals entered pursuant to this subsection or subsection (a), the court of appeals may assess the penalties provided in Section 17, in addition to invoking any other available remedies.*

PROCEDURES TO COUNTERACT IMMINENT DANGERS

In certain hazardous situations, an employer can be ordered to shut down his entire operation, or part of it, practically at once. A federal court can issue such an order whenever it finds a danger that could reasonably be expected to cause death or serious physical harm immediately or before the danger can be removed through the regular enforcement procedure.

Examples of such hazardous conditions against which shutdown orders have been issued are the presence of methane gas in a tunnel worksite and an improperly shored trench that might have caved in at any time.

What is the procedure? Should employees or their representative feel there is an imminent danger, they can ask for a special inspection. On either a special or routine inspection, if the inspector feels that there is an immediate danger, he will tell both employer and employees about the danger and that he is recommending a shutdown proceeding. The Labor Department can then bring the case to court.

Procedure by inspector

If an inspector finds an imminent danger to employees exists, he will inform them and try to get the employer to remove employees from the dangerous area. If the compliance procedure fails, the inspector will notify the area director immediately (by phone, if possible) and tell the employer and employees of his recommendation. The inspector has no authority to order a shutdown. This can only be done by a court order after suit is brought by the Department of Labor.

The shutdown order

The court order may be limited. For example, it can forbid an employer to use a particular machine or substance, or it can forbid him to let employees work under certain conditions or in certain areas. The employer can be ordered to remove the danger as well. Or he could be ordered to shut down entirely until the danger is removed.

In some cases, however, an employer may be permitted to use a limited number of people to maintain a continuous-process operation. But the continuous-process operation would be allowed only if it could be done safely. Also allowed would be the persons necessary to remove the danger, and persons necessary to allow a shutdown to be accomplished safely and in an orderly manner.

Hearings

A court can issue a shutdown order (temporary restraining order) for up to five days without a hearing. To extend the order beyond five days, a hearing is necessary. Normally a hearing date will be set when the temporary restraining order is issued.

Employees can force the issue

The Labor Department may decide not to ask for a shutdown order. In that event, any employee who is affected, or his representative, can bring the department to court; upon proof that the department's decision was "arbitrary or capricious," the court can order the Labor Department to sue for a shutdown.

"AS STATED"
Sec. 13. "THE ACT"

Sec. 13.(a) *The United States district courts shall have jurisdiction, upon petition of the Secretary, to restrain any conditions or practices in any place of employment which are such that a danger exists which could reasonably be expected to cause death or serious physical harm immediately or before the imminence of such danger can be eliminated through the enforcement procedures otherwise provided by this Act. Any order issued under this section may require such steps to be taken as may be necessary to avoid, correct, or remove such imminent danger and prohibit the employment or presence of any individual in locations or under conditions where such imminent danger exists, except individuals whose presence is necessary to avoid, correct, or remove such imminent danger or to maintain the capacity of a continuous process operation to resume normal operations without a complete cessation of operations, or where a cessation of operations is necessary, to permit such to be accomplished in a safe and orderly manner.*

Sec. 13.(b) *Upon the filing of any such petition the district court shall have jurisdiction to grant such injunctive relief or temporary restraining order pending the outcome of an enforcement proceeding pursuant to this Act. The proceeding shall be as provided by Rule 65 of the Federal Rules, Civil Procedure, except that no temporary restraining order issued without notice shall be effective for a period longer than five days.*

Sec. 13.(c) *Whenever and as soon as an inspector concludes that conditions or practices described in subsection (a) exist in any place of employment, he shall inform the affected employees and employers of the danger and that he is recommending to the Secretary that relief be sought.*

Sec. 13.(d) *If the Secretary arbitrarily or capriciously fails to seek relief under this section, any employee who may be injured by reason of such failure, or the representative of such employees, might bring an action against the Secretary in the United States district court for the district in which the imminent danger is alleged to exist or the employer has its principal office, or for the District of Columbia, for a writ of mandamus to compel the Secretary to seek such an order and for such further relief as may be appropriate.*

REPRESENTATION IN CIVIL LITIGATION

The solicitor of labor may represent the secretary in any civil litigation.

"AS STATED"
Sec. 14. "THE ACT"
Sec. 14. *Except as provided in Section 518(a) of Title 28, United States Code, relating to litigation before the Supreme Court, the Solicitor of Labor may appear for and represent the Secretary in any civil litigation brought under this Act but all such litigation shall be subject to the direction and control of the Attorney General.*

7

VARIANCES

VARIANCES

An employer may apply to the secretary of labor for a *temporary variance* from a standard if he is unable to comply because of the unavailability of materials, equipment, or personnel to make changes within the required time. He may also apply for a *permanent variance* from a standard if he can prove that his facilities or method of operation provide protection for employees that is at least as effective as that required by the standard.

Not retroactive: All variances granted under the act are only effective in the future; thus an employer cannot get a variance to relieve him of a penalty if he has already been cited for a violation of the act. And the secretary will not even consider the variance application if the employer has been cited for a specific violation of that standard until after the abatement period for the violation and any proceeding before the Review Commission is completed.

Provisions and procedures for variances are as follows:

TEMPORARY VARIANCE ORDER

You may apply for a temporary variance for the following reasons:

1 / You are unable to comply with a standard because of the unavailability of personnel, material, or equipment needed to come into compliance in time.

2 / The necessary construction or alteration of facilities cannot be completed in time.

However, you must be taking all available steps to safeguard employees, and you must have an effective program for coming into compliance.

Any temporary variance must prescribe the practices, means, methods, operations, and processes that will be used while the temporary variance is in effect, and must state in detail your program for coming into compliance.

A temporary variance may only be granted after notice to employees and an opportunity for a hearing, except that one interim order may be granted to be in effect (for a maximum of 180 days) until a decision is made based on the hearing.

Temporary variances are good for only as long as needed to achieve compliance but for not more than one year (whichever is shorter). Temporary variances may be renewed not more than twice. Application for renewal must be filed ninety days prior to the expiration date.

"AS STATED"
Sec. 6. "THE ACT"
Sec. 6.(b)(6)(A) *Any employer may apply to the Secretary for a temporary order granting a variance from a standard or any provision thereof promulgated under this section. Such temporary order shall be granted only if the employer files an application which meets the requirements of clause (B) and establishes that (i) he is unable to comply with a standard by its effective date because of unavailability of professional or technical personnel or of materials and equipment needed to come into compliance with the standard or because necessary construction or alteration of facilities cannot be completed by the effective date, (ii) he is taking all available steps to safeguard his employees against the hazards covered by the standard, and (iii) he has an effective program for coming into compliance with the standard as quickly as practicable. Any temporary order issued under this paragraph shall prescribe the practices, means, methods, operations, and processes which the employer must adopt and use while the order is in effect and state in detail his program for coming into compliance with the standard. Such a temporary order may be granted only after notice to employees and an opportunity for a hearing: Provided, That the Secretary may issue one interim order to be effective until a decision*

is made on the basis of the hearing. No temporary order may be in effect for longer than the period needed by the employer to achieve compliance with the standard or one year, whichever is shorter, except that such an order may be renewed not more than twice (I) so long as the requirements of this paragraph are met and (II) if an application for renewal is filed at least 90 days prior to the expiration date of the order. No interim renewal of an order may remain in effect for longer than 180 days.

TEMPORARY VARIANCE APPLICATION

An application for a temporary variance must contain the following:

"AS STATED"
Sec. 6. "THE ACT"
Sec. 6.(b)(6)(B) *An application for a temporary order under this paragraph (6) shall contain:*

(i) a specification of the standard or portion thereof from which the employer seeks a variance,

(ii) a representation by the employer, supported by representations from qualified persons having firsthand knowledge of the facts represented, that he is unable to comply with the standard or portion thereof and a detailed statement of the reasons therefor,

(iii) a statement of the steps he has taken and will take (with specific dates) to protect employees against the hazard covered by the standard.

(iv) a statement of when he expects to be able to comply with the standard and what steps he has taken and what steps he will take (with dates specified) to come into compliance with the standard, and

(v) a certification that he has informed his employees of the application by giving a copy thereof to their authorized representative, posting a statement giving a summary of

the application and specifying where a copy may be examined at the place or places where notices to employees are normally posted, and by other appropriate means.

A description of how employees have been informed shall be contained in the certification. The information to employees shall also inform them of their right to petition the Secretary for a hearing.

The last provision, notice to employees, is very important. Your employees have a right to know that you are applying for a variance, and it is your responsibility to tell them. Your employees are also entitled to ask the secretary for a hearing on the application.

VARIANCE FOR EXPERIMENTAL WORK

You can be granted a variance to permit employees to participate in an experiment designed to demonstrate new or improved techniques relating to safety and health.

"AS STATED"
Sec. 6. "THE ACT"
Sec. 6.(b)(6)(C) *The Secretary is authorized to grant a variance from any standard or portion thereof whenever he determines, or the Secretary of Health, Education, and Welfare certifies, that such variance is necessary to permit an employer to participate in an experiment approved by him or the Secretary of Health, Education, and Welfare designed to demonstrate or validate new and improved techniques to safeguard the health or safety of workers.*

PERMANENT VARIANCE

You can be granted a permanent variance if you can prove that the conditions, practices, means, methods, operations, or processes used will provide as safe and healthful conditions as would prevail if you complied with the standard.

"AS STATED"
Sec. 6. "THE ACT"
Sec. 6.(d) *Any affected employer may apply to the Secretary for a rule or order for a variance from a standard promulgated under this section. Affected employees shall be given notice of each such application and an opportunity to participate in a hearing. The Secretary shall issue such rule or order if he determines on the record, after opportunity for an inspection where appropriate and a hearing, that the proponent of the variance has demonstrated by a preponderance of the evidence that the conditions, practices, means, methods, operations, or processes used or proposed to be used by an employer will provide employment and places of employment to his employees which are as safe and healthful as those which would prevail if he complied with the standard. The rule or order so issued shall prescribe the conditions the employer must maintain, and the practices, means, methods, operations, and processes which he must adopt and utilize to the extent they differ from the standard in question. Such a rule or order may be modified or revoked upon application by an employer, employees, or by the Secretary on his own motion, in the manner prescribed for its issuance under this subsection at any time after six months from its issuance.*

VARIATIONS, TOLERANCES, AND EXEMPTIONS

OSHA regulations shall not seriously impair national defense.

"AS STATED"
Sec. 16. "THE ACT"
Sec. 16. *The Secretary, on the record, after notice and opportunity for a hearing may provide such reasonable limitations and may make such rules and regulations allowing reasonable variations, tolerances, and exemptions to and from any or all provisions of this Act as he may find necessary and proper to avoid serious impairment of the national*

defense. Such action shall not be in effect for more than six months without notification to affected employees and an opportunity being afforded for a hearing.

8

DISCRIMINATION PROTECTION

DISCRIMINATION PROTECTION

ANY PERSON IS PROTECTED FROM DISCRIMINATION FOR ASSERTING HIS RIGHTS UNDER THE LAW

The act provides that "No person shall discharge or in any manner discriminate against any employee because such employee":

1 / Filed a complaint.

2 / Instituted or caused to be instituted any proceeding under or related to the act.

3 / Testified or is about to testify in any such proceeding.

4 / Exercised on behalf of himself or others any right given by the act.

The Labor Department has issued an interpretation of this provision, setting out procedures to be followed by employees claiming discrimination.

Who is covered by the provision

The act bans discrimination by any person, so it is not limited to discrimination by an employer against his employee. "Persons" is defined to include "individuals, partnerships, associations, corporations, business trusts, legal representatives, or any group of persons." The law therefore bans discrimination not only by employers but by labor unions, employment agencies, or anyone else in a position to discriminate.

Discrimination against any employee is banned, not just employees of the employer being charged with discrimination. The term "employee" is broad enough to include applicants for employment and former employees.

What is protected

The act protects employees' rights to make complaints, start proceedings, give testimony, or exercise any rights under the act. These activities are not limited to formal judicial proceedings, but include such actions as requesting an inspection, contest of abatement dates, starting proceedings to set a standard, making statements during investigations, requesting information from OSHA, and any other action "under or related to the Act."

"Related to the Act"

An example of an activity protected because it is "related to the Act" would be the filing of a complaint with another federal agency (or state agency) with authority to regulate or investigate occupational health and safety conditions; this would apply only to complaints about workplace safety and health, not just general public safety and health.

Test for discrimination

If a protected activity was a substantial reason for the discipline, discharge, or other adverse action, or if the action would not have taken place "but for" the protected activity, then the act has been violated.

Employers may be required to rehire or reinstate an employee to his former position with back pay.

DISCRIMINATION PROHIBITED

"AS STATED"
Sec. 11. "THE ACT"
Sec. 11.(c)(1) *No person shall discharge or in any manner discriminate against any employee because such employee has filed any complaint or instituted or caused to be insti-*

tuted any proceeding under or related to this Act or has testified or is about to testify in any such proceeding or because of the exercise by such employee on behalf of himself or others of any right afforded by this Act.

EMPLOYEE'S COMPLAINT

Any employee who believes that he has been discharged or discriminated against may file a complaint. Contact the area director (see the Directory at the back of this manual) for filing your complaint. If found guilty, the employer can be made to provide appropriate relief, including rehiring and reinstatement with back pay.

"AS STATED"
Sec. 11. "THE ACT"
Sec. 11.(c)(2) *Any employee who believes that he has been discharged or otherwise discriminated against by any person in violation of this subsection may, within thirty days after such violation occurs, file a complaint with the Secretary alleging such discrimination. Upon receipt of such complaint, the Secretary shall cause such investigation to be made as he deems appropriate. If upon such investigation, the Secretary determines that the provisions of this subsection have been violated, he shall bring an action in any appropriate United States district court against such person. In any such action the United States district courts shall have jurisdiction, for cause shown to restrain violations of paragraph (1) of this subsection and order all appropriate relief including rehiring or reinstatement of the employee to his former position with back pay.*

A complaint for discrimination under the act must be filed by the employee or his authorized representative with the OSHA area director within thirty days of the alleged discrimination. The thirty-day period can be extended if the employee (within thirty days) files a complaint with another agency on the same subject, or if he starts grievance-arbitration proceedings, or if the employer has concealed from or misled the employee about the grounds for discharge or other action, or on other recognized equitable grounds.

Once a complaint is filed, an employee cannot automatically withdraw it. He can request withdrawal, but the secretary of labor does not have to consent, if he finds the public interest is involved.

NOTIFICATION OF DETERMINATION

Employees filing a complaint shall be notified within ninety days of the ruling.

> "AS STATED"
> Sec. 11. "THE ACT"
> Sec. 11.(c)(3) *Within 90 days of the receipt of a complaint filed under this subsection the Secretary shall notify the complainant of his determination under paragraph 2 of this subsection.*

CAN AN EMPLOYEE SUE?

If the secretary does not find any discrimination, or if he declines to sue, can an employee start an action in the district court on his own? The act does not specifically give him such a right. However, there has been no official determination yet.

OTHER ACTIONS BY EMPLOYEES

Suppose that an employee complains of discrimination to another agency, such as the National Labor Relations Board, or starts grievance-arbitration proceedings under a union contract. The secretary of labor is not bound by the results of the other proceeding. However, he will defer to the result if it is clear that "those proceedings dealt adequately with all factual issues, that the proceedings were fair, regular, and free of procedural infirmities, and that the outcome was not repugnant to the purpose and policy of the Act." If the other proceedings are dismissed without a hearing, the secretary will not consider himself bound.

If an employee files a complaint with the secretary of labor and also starts other proceedings, the secretary may, if he chooses, postpone his decision if it appears that the other proceedings will dispose of the matter. However, he is still not bound by the result unless all the conditions mentioned above have been met.

Strikes and Work Stoppages

OSHA says there is no right under the act for employees to walk off the job because of "potential unsafe conditions at the workplace." But there is one exception. If the employee, by performing his assigned work, would "subject himself to serious injury or death from a hazardous condition at the workplace," and "with no reasonable alternative, refuses in good faith to expose himself to the dangerous condition, he would be protected against subsequent discrimination." Not only must it appear to a reasonable person that there is serious danger but also "that there is insufficient time, due to the urgency of the situation, to eliminate the danger through resort to regular statutory enforcement channels."

Strikes: Other labor laws make only one specific reference to safety, but that is a vitally important one: 502 of the Taft–Hartley Act says "the quitting of labor by an employee or employees in good faith because of abnormally dangerous conditions for work" is not to be considered a strike under Taft–Hartley.

Under this provision, the Court of Appeals for the Third Circuit refused to grant an injunction against a safety strike at a coal mine, even though the union contract had a no-strike clause and a provision for settling disputes by arbitration. (Under these conditions, courts will normally grant injunctions.) But the court said that even the federal labor policy in favor of arbitration and voluntary peaceful settlement of disputes would not justify forcing employees to accept an arbitrator's decision when they fear their lives or health are at stake.

On the other hand the Court of Appeals for the Eighth Circuit did grant an injunction, because the union contract contained not just a general grievance-arbitration clause, but specific provisions for dealing with safety disputes, including relieving or reassigning employees who felt endangered. However, the court put strict conditions on the injunction, requiring the company to take action to ensure employee safety, and insisting on swift arbitration.

The effect of the OSHA regulations on other labor relations laws (and vice versa) remains to be decided. It seems likely, though, that courts will be less willing to grant injunctions in safety cases than in cases of other labor disputes and will be stricter about their conditions.

Collective Bargaining

Before passage of OSHA, most labor contracts contained only general clauses on maintaining a safe place to work. Since the act, collective bargaining on safety issues has been much more extensive and contract clauses more detailed. Some bargaining issues arise directly from the act (e.g., union demands for pay for walk-around inspection time), but many are just due to the increasing safety consciousness of both management and labor.

Here, for example, are some objectives one union told its negotiators to push for at contract time:

1 / Protect the worker's legitimate right to leave any job he considers hazardous or unhealthy.

2 / Set up a joint safety committee with the authority to recommend rules after investigating unsafe procedures or hazardous conditions.

3 / Establish a formal mechanism to resolve safety–health grievances. Treat any health or safety grievance as a top-priority item and resolve it immediately.

4 / Ask for a safety program that would include assigning a union safety chairman to work full time on health and safety with a salary commensurate with his previous pay level.

5 / Give all members a complete physical at least once a year. The

minimum requirements should be tests of vision, hearing, and lung air capacity and a blood pressure check.

One reason management has to be concerned with collective bargaining on safety and the operation of grievance-arbitration clauses in union contracts is that the act provides penalties for employers but not for employees; the framers of the law apparently preferred to leave discipline of employees to traditional labor relations methods.

A number of arbitration awards have indicated that, although management definitely has the right to make and enforce safety rules, the rules have to be administered equitably. For instance, in a number of cases, employers have done little or nothing about enforcing safety rules, then suddenly have imposed harsh discipline or even discharge. The arbitrators have almost invariably upheld the right to discipline, but have reduced the penalties drastically on equitable grounds.

The moral: be consistent. Make rules as definite as possible; then enforce them evenhandedly as to all employees without arbitrary fluctuations between overlaxity and overharshness.

9

OSHA MANDATORY ACCIDENT REPORTING

OSHA MANDATORY ACCIDENT REPORTING

Mandatory requirements for reporting incidents of occupational (work-related) deaths, injuries, or illnesses are as follows:

> • **CAUTION** • OSHA reporting requirements are separate and different from OSHA recordkeeping requirements. For OSHA recordkeeping requirements, see Chapter 10.

NOTIFICATION

Only certain occupational (work-related) incidents are required to be reported. Non-work-related incidents are not required to be reported. The OSHA area director must be notified within forty-eight hours following the occurrence of any of the following incidents:

Deaths: Every fatality must be reported as noted above.

Injuries: Every occurrence resulting in five or more hospitalized cases must be reported as noted above.

Illnesses: Every occurrence resulting in five or more hospitalized cases must be reported as noted above.

METHOD OF NOTIFICATION

Send a telegram to the OSHA area director at the nearest OSHA regional office (see the Directory at the back of this manual) with notification of the occurrence as soon as possible (within forty-eight hours) after the reportable incident occurs. Notification may be made orally or in writing, but a written or telegraphic report is preferable.

PROVISIONS FOR ACCIDENT REPORTING

There are provisions in the law to report work-related deaths, injuries, and illnesses.

"AS STATED"
Sec. 8. "THE ACT"
Sec. 8.(c)(2) *The Secretary, in cooperation with the Secretary of Health, Education, and Welfare, shall prescribe regulations requiring employers to maintain accurate records of, and* to make periodic reports on, work-related deaths, injuries, and illnesses *other than minor injuries requiring only first aid treatment and which do not involve medical treatment, loss of consciousness, restriction of work or motion, or transfer to another job.*

10

OSHA RECORDKEEPING

OSHA RECORDKEEPING

Mandatory recordkeeping requirements are as follows:

EVERY EMPLOYER

Every employer who has or has had eight or more employees at any one time during the calendar year immediately preceding the current calendar year *must* maintain three basic records.

> • **NOTE** • For fiscal 1975, Congress has extended the recordkeeping exemptions to employers with ten or fewer employees instead of the present cutoff of seven or fewer employees. Thus, during 1975, if you have fewer than eleven employees you are not required to keep OSHA Forms 100, 101, and 102.

The three basic records are OSHA Form 100 (the Log), Form 101 (Supplementary Record), and Form 102 (Annual Summary). These records must be kept current, readily available in each establishment without delay and at reasonable times for examination by representatives of the Department of Labor and/or the Department of Health, Education, and Welfare. The records must be maintained for a period of five years following the end of the calendar year to which they relate.

All records are the employer's records. There are no requirements to send in or submit these records to any agency of the federal government or others unless specifically requested to do so. *Note:* OSHA compliance officers making inspections will request to see these records.

PETITIONS FOR RECORDKEEPING VARIANCES

Any employer who wants to keep records in a different manner from the regulations may submit a petition outlining the desired procedure to the regional director of the Bureau of Labor Statistics for the establishment's area. Affected employees or their representatives must be given a copy of the petition and a summary of the petition must be posted. Employees then have ten working days within which to present written data, views, or arguments on the petition.

SPECIAL OSHA SURVEY

In addition to maintaining OSHA Forms 100, 101, and 102, if an employer with any number of employees is notified by the proper authorities that the firm has been selected to participate in a statistical survey, that employer must maintain a log of all occupational injuries or illnesses for the prescribed year. In this case, the notified employer will be provided with instructions and OSHA Survey Form 103 (see reduced-size facsimile at the end of the chapter).

INFORMATION TO BE RECORDED

Required recordable occupational (work-related) deaths, injuries, or illnesses are as follows:

Fatalities

All occupational (work-related) deaths, regardless of time between a related injury and death, or regardless of the length of a death-related illness, shall be recorded.

Lost workday cases

Cases other than fatalities that result in lost workdays (except the day of the mishap) shall be recorded.

Nonfatal cases without lost workdays

1 / All cases that result in transfer to another job or termination of employment shall be recorded.
2 / All cases requiring medical treatment (as hereinafter defined) shall be recorded.

3 / All cases that involve loss of consciousness or restriction of work or motion shall be recorded.

4 / All cases of diagnosed occupational (work-related) illnesses that are reported to the employer but are not classified as fatalities or lost workday cases shall be recorded.

Medical Treatment

Medical treatment includes treatment administered by a physician or by registered professional personnel under the standing orders of a physician. Medical treatment does not include first aid treatment (one-time treatment and subsequent observation of minor scratches, cuts, burns, splinters, etc., that do not ordinarily require medical care) even though provided by a physician or registered personnel.

RECORDKEEPING REQUIREMENTS CHART

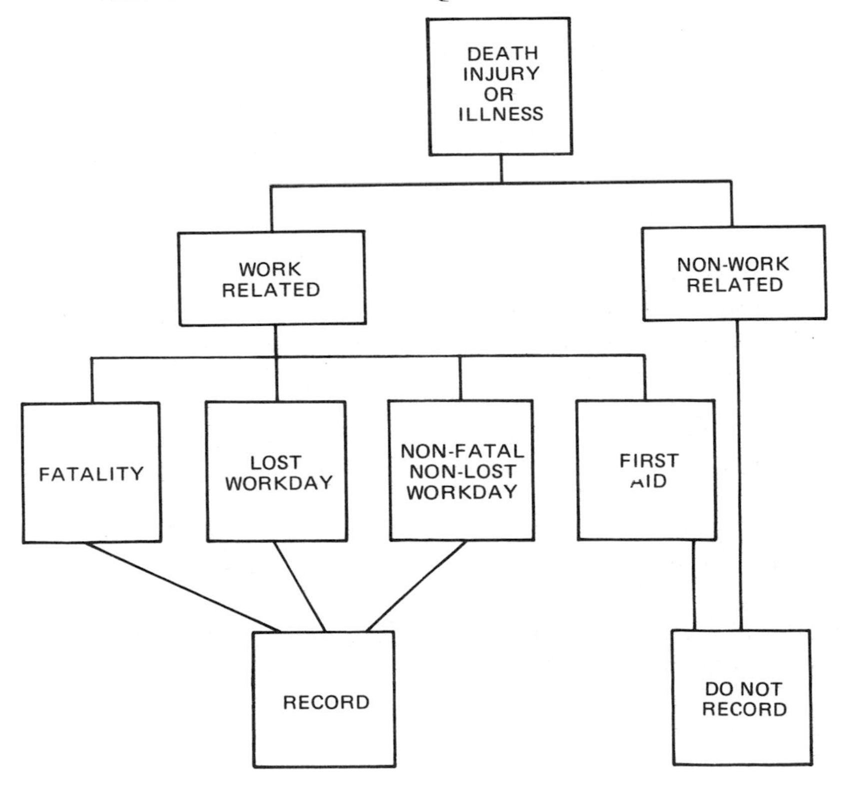

REQUIRED FORMS

Forms may be obtained from the nearest OSHA office. (See the Directory at the back of this manual.) Instructions are printed on the reverse side of each form. Each form should have the year to which it relates noted on the form. Simply note at the top of the form "For Calendar Year 19______."

OSHA Form 100

OSHA Form 100 (see reduced-size facsimile at the end of the chapter) is a log of all occupational injuries and illnesses that are required to be recorded.

Maintaining the log

Form 100 (the log) is required to be maintained at each of the employer's establishments, kept readily available and current to within six days.

> • **EXCEPTION** • Form 100 (the log) may be maintained at another location (and/or by means of data-processing equipment) provided that (1) the information is kept current to within six days, (2) is readily available at the other location, and (3) a copy of the log, current to within forty-five days, is readily available at each of the employer's establishments to which the log applies.

Initialing the log

Individual entries on Form 100 (the log) no longer need to be initialed as originally required.

OSHA Form 101

OSHA Form 101 (see reduced-size facsimile at the end of this chapter) or its equivalent containing the same information is a required supplementary record of occupational injuries and illnesses which must be completed for each recordable incident.

Substitute forms

Workmen's compensation forms, standard insurance forms, other reports, or a separate plain sheet of paper are acceptable substitutes that may be used in lieu of OSHA Form 101, provided the required facts are included thereon.

Deadline

The supplementary Form 101 (or allowable substitute) record must be at each establishment within six working days after the employer knows that a recordable case has occurred.

OSHA Form 102

OSHA Form 102 (see reduced-size facsimile at the end of this chapter) is a required annual summary of all occupational injuries and illnesses.

Year-end preparation

Every employer with eight or more employees as noted must prepare an annual summary (Form 102) of the occupational injury and illness experience of the employees for each of the employer's establishments at the end of each calendar year within one month following the end of that year. OSHA Form 102 must be used unless the employer petitions for and is granted a variation.

Certification

The person responsible for supervising the preparation of the annual summary (Form 102) must certify that it is true and complete by signing it in the lower-right corner.

Incident-free years

Should any year yield no occurrences of occupational injuries, illness, or deaths, simply note on that year's forms "No Occurrences." However, all occurrences must be recorded as required if they do in fact occur.

Posting requirements

The completed annual summary (Form 102) must be posted no later than February 1 and remain in place until March 1 of each year in a prominent place accessible to the employees in each establishment.

> • **EXCEPTION** • For employees who do not report to work at a single establishment, or who do not report to any fixed establishment on a regular basis, employers shall satisfy the posting requirements by presenting or mailing a copy of the completed annual summary (Form 102) during the month of February to all such employees who receive pay during that month.

RECORDS FOR EMPLOYEES NOT IN FIXED ESTABLISHMENTS

The following regulations shall apply to employees who are engaged in physically dispersed operations such as occur in construction, installation, repair, or service activities, and who do not report to any fixed establishment on a regular basis, but who are subject to common supervision (field superintendent, field supervisor, etc.).

Records covering such employees may be maintained in an established central place provided that the address and telephone number of the central recordkeeping place are available at the worksite and that during business hours personnel are available at the central place to provide information from the records by telephone or mail.

> • **CAUTION** • This regulation does not automatically apply to all construction, installation, repair, or service activities. If in doubt about applicability to your operation, contact the OSHA area director at the nearest OSHA office. (See the Directory at the back of this manual.)

PROVISIONS FOR RECORDKEEPING

The provisions within the law for recordkeeping are as follows:

"AS STATED"
Sec. 8. "THE ACT"
Sec. 8.(c)(1) *Each employer shall make, keep and preserve, and make available to the Secretary or the Secretary of Health, Education, and Welfare, such records regarding his activities relating to this Act as the Secretary, in cooperation with the Secretary of Health, Education, and Welfare, may prescribe by regulation as necessary or appropriate for the enforcement of this Act or for developing information regarding the causes and prevention of occupational accidents and illnesses. In order to carry out the provisions of this paragraph such regulations may include provisions requiring employers to conduct periodic inspections. The Secretary shall also issue regulations requiring that employers, through posting of notices or other appropriate means, keep their employees informed of their protections and obligations under this Act, including the provisions of applicable standards.*

Sec. 8. "THE ACT"
Sec. 8.(c)(2) *The Secretary, in cooperation with the Secretary of Health, Education, and Welfare, shall prescribe*

regulations requiring employers to maintain accurate records of, and to make periodic reports on, work-related deaths, injuries, and illnesses other than minor injuries requiring only first aid treatment and which do not involve medical treatment, loss of consciousness, restriction of work or motion, or transfer to another job.

RECORDS FOR EXPOSURE TO TOXIC MATERIALS OR HARMFUL PHYSICAL AGENTS

Employers are required to maintain separate records of employees exposed to toxic materials or harmful physical agents that are required to be monitored. Each employer must notify any employee who has or is being exposed. Employees shall have an opportunity to observe monitoring. Each employee, as well as past employees, shall have access to records that indicate their own exposure.

"AS STATED"
Sec. 8. "THE ACT"
Sec. 8.(c)(3) *The Secretary, in cooperation with the Secretary of Health, Education, and Welfare, shall issue regulations requiring employers to maintain accurate records of employee exposures to potentially toxic materials or harmful physical agents which are required to be monitored or measured under Section 6. Such regulations shall provide employees or their representatives with an opportunity to observe such monitoring or measuring, and to have access to the records thereof. Such regulations shall also make appropriate provision for each employee or former employee to have access to such records as will indicate his own exposure to toxic materials or harmful physical agents. Each employer shall promptly notify any employee who has been or is being exposed to toxic materials or harmful physical agents in concentrations or at levels which exceed those prescribed by an applicable occupational safety and health standard promulgated under Section 6, and shall inform any employee who is being thus exposed of the corrective action being taken.*

MINIMUM BURDEN PROVISION

The law provides that recordkeeping shall be obtained with a minimum burden upon employers, especially small businesses.

"AS STATED"
Sec. 8. "THE ACT"
Sec. 8.(d) *Any information obtained by the Secretary, the Secretary of Health, Education, and Welfare, or a State agency under this Act shall be obtained with a minimum burden upon employers, especially those operating small businesses. Unnecessary duplication of efforts in obtaining information shall be reduced to the maximum extent feasible.*

OSHA NO. 100

LOG OF OCCUPATIONAL INJURIES AND ILLNESSES

CASE OR FILE NUMBER (1)	DATE OF INJURY OR ONSET OF ILLNESS Mo./day/yr. (2)	EMPLOYEE'S NAME (First name or initial, middle initial, last name) (3)	OCCUPATION (Enter regular job title, not activity employee was performing when injured or at onset of illness.) (4)	DEPARTMENT (Enter department in which the employee is regularly employed.) (5)

Company Name ____________________
Establishment Name ____________________
Establishment Address ____________________

NOTE: This is NOT a report form. Keep it in the establishment for 5 years.

Reduced Size Facsimile

RECORDABLE CASES: You are required to record information about: every occupational death; every nonfatal occupational illness; and those nonfatal occupational injuries which involve one or more of the following: loss of consciousness, restriction of work or motion, transfer to another job, or medical treatment (other than first aid).
More complete definitions appear on the other side of this form.

Form Approved
OMB No. 44R 1453

DESCRIPTION OF INJURY OR ILLNESS		EXTENT OF AND OUTCOME OF CASES					
			LOST WORKDAY CASES			NONFATAL CASES WITHOUT LOST WORKDAYS	TERMINATIONS OR PERMANENT TRANSFERS
				LOST WORKDAYS			
Nature of Injury or Illness and Part(s) of Body Affected (Typical entries for this column might be: Amputation of 1st joint right forefinger Strain of lower back Contact dermatitis on both hands Electrocution—body)	Injury or Illness Code See codes at bottom of page.	DEATHS (Enter date of death.) Mo./day/yr.	Enter a check if case involved lost workdays.	Enter number of days AWAY FROM WORK due to injury or illness.	Enter number of days of RESTRICTED WORK ACTIVITY due to injury or illness.	(Enter a check if no entry was made in columns 8 or 9 but the case is recordable, as defined above.)	(Enter a check if the entry in columns 9 or 10 represented a termination or permanent transfer.)
(6)	(7)	(8)	(9)	(9A)	(9B)	(10)	(11)

Injury Code
10 All occupational injuries
Illness Codes

21 Occupational skin diseases or disorders
22 Dust diseases of the lungs (pneumoconioses)
23 Respiratory conditions due to toxic agents
24 Poisoning (systemic effects of toxic materials)
25 Disorders due to physical agents (other than toxic materials)
26 Disorders associated with repeated trauma
29 All other occupational illnesses

Reduced Size Facsimile

LOG OF OCCUPATIONAL INJURIES AND ILLNESSES

Each employer who is subject to the recordkeeping requirements of the Occupational Safety and Health Act of 1970 must maintain for each establishment a log of all recordable occupational injuries and illnesses. This form (OSHA No. 100) may be used for that purpose. A substitute for the OSHA No. 100 is acceptable if it is as detailed, easily readable and understandable as the OSHA No. 100.

Each recordable occupational injury and occupational illness must be promptly entered on the log. Logs must be kept current and retained for five (5) years following the end of the calendar year to which they relate. Logs must be available (normally at the establishment) for inspection and copying by representatives of the Department of Labor, or the Department of Health, Education and Welfare, or states accorded jurisdiction under the act.

INSTRUCTIONS FOR COMPLETING LOG OF OCCUPATIONAL INJURIES AND ILLNESSES

Column 1—CASE OR FILE NUMBER

Enter a number which will facilitate comparison with supplementary records. Any series of nonduplicating numbers may be used.

Column 2—DATE OF INJURY OR ONSET OF ILLNESS

For occupational injuries enter the date of the work accident which resulted in injury. For occupational illnesses enter the date of initial diagnosis of illness, or, if absence from work occurred before diagnosis, enter the first day of the absence attributable to the illness which was later diagnosed or recognized.

Column 3—EMPLOYEE'S NAME

Column 4—OCCUPATION

Enter regular job title, not the specific activity being performed at time of injury or illness. In the absence of a formal occupational title, enter a brief description of the duties of the employee.

Column 5—DEPARTMENT

Enter the name of the department or division in which the injured person is regularly employed, even though temporarily working in another department at the time of injury or illness. In the absence of formal department titles, enter a brief description of normal workplace to which employee is assigned.

Column 6—NATURE OF INJURY OR ILLNESS AND PART(S) OF BODY AFFECTED

Enter a brief description of the injury or illness and indicate the part or parts of body affected. Where entire body is affected, the entry "body" can be used.

Column 7—INJURY OR ILLNESS CODE

Enter the one code which most accurately describes the case. A list of the codes appears at the bottom of the log. A more complete description of recordable occupational injuries and illnesses appears in "DEFINITIONS."

Column 8—DEATHS

If the occupational injury or illness resulted in death, enter date of death.

Column 9—LOST WORKDAY CASES

Enter a check for each case which involves days away from work, or days of restricted work activity, or both. Each lost workday case also requires an entry in column 9A or column 9B, or both.

Column 9A—LOST WORKDAYS—DAYS AWAY FROM WORK

Enter the number of workdays (consecutive or not) on which the employee would have worked but could not because of occupational injury or illness. The number of lost workdays should not include the day of injury or onset of illness or any days on which the employee would not have worked even though able to work.

• **NOTE** • For employees not having a regularly scheduled shift, i.e., certain truck drivers, construction workers, farm labor, casual labor, part-time employees, etc., it may be necessary to estimate the number of lost workdays. Estimates of lost workdays shall be based on prior work history of the employee AND days worked by employees, not ill or injured, working in the department and/or occupation of the ill or injured employee.

Column 9B—LOST WORKDAYS—DAYS OF RESTRICTED WORK ACTIVITY

Enter the number of workdays (consecutive or not) on which because of injury or illness:

1) the employee was assigned to another job on a temporary basis, or
2) the employee worked at a permanent job less than full time, or
3) the employee worked at a permanently assigned job but could not perform all duties normally connected with it.

The number of lost workdays should not include the day of injury or onset of illness or any days on which the employee would not have worked even though able to work.

Column 10—NONFATAL CASES WITHOUT LOST WORKDAYS

Enter a check for any recordable case which does not involve a fatality or lost workdays.

Column 11—TERMINATIONS OR PERMANENT TRANSFERS

Enter a check if the entry in columns 9 or 10 represented a termination of employment or permanent transfer.

CHANGES IN EXTENT OF OR OUTCOME OF INJURY OR ILLNESS

If, during the 5-year period the log must be retained, there is a change in a case which affects entries in columns 9 or 10, the first entry should be lined out and a new entry made. For example, if an injured employee at first required only medical treatment but later lost workdays, the check in column 10 should be lined out, a check entered in column 9, and the number of lost workdays entered in columns 9A and/or 9B.

In another example, if an employee with an occupational illness lost workdays, returned to work, and then died of the illness, the entries in columns 9, 9A, and/or 9B should be lined out and the date of death entered in column 8.

The entire entry for a case should be lined out if the case is later found to be nonrecordable. Examples are: A case which is later determined not to be work related, or a case which was initially thought to involve medical treatment but later was determined to have involved only first aid.

DEFINITIONS

RECORDABLE OCCUPATIONAL INJURIES AND ILLNESSES are:

1) OCCUPATIONAL DEATHS, regardless of the time between injury and death, or the length of the illness; or

2) OCCUPATIONAL ILLNESSES; or

3) OCCUPATIONAL INJURIES which involve one or more of the following: loss of consciousness, restriction of work or motion, transfer to another job, or medical treatment (other than first aid).

> • **NOTE** • Any case which involves lost workdays must be recorded since it always involves one or more of the criteria for recordability.

OCCUPATIONAL INJURY is any injury such as a cut, fracture, sprain, amputation, etc., which results from a work accident or from an exposure involving a single incident in the work environment.

> • **NOTE** • Conditions resulting from animal bites, such as insect or snake bites, or from one-time exposure to chemicals are considered to be injuries.

OCCUPATIONAL ILLNESS of an employee is any abnormal condition or disorder, other than one resulting from an occupational injury, caused by exposure to environmental factors associated with employment. It includes acute and chronic illnesses or diseases which may be caused by inhalation, absorption, ingestion, or direct contact.

The following listing gives the categories of occupational illnesses and disorders that will be utilized for the purpose of classifying recordable illnesses. The identifying codes are those to be used in column 7 of the log. For purposes of information, examples of each category are given. These are typical examples, however, and are not to be considered to be the complete listing of the types of illnesses and disorders that are to be counted under each category.

(21) Occupational Skin Diseases or Disorders
Examples: Contact dermatitis, eczema, or rash caused by primary irritants and sensitizers or poisonous plants; oil acne; chrome ulcers; chemical burns or inflammations; etc.

(22) Dust Diseases of the Lungs (Pneumoconioses)
Examples: Silicosis, asbestosis, coal worker's pneumoconiosis, byssinosis, and other pneumoconioses.

(23) Respiratory Conditions Due to Toxic Agents
Examples: Pneumonitis, pharyngitis, rhinitis or acute congestion due to chemicals, dusts, gases, or fumes; farmer's lung; etc.

(24) Poisoning (Systemic Effects of Toxic Materials)

Examples: Poisoning by lead, mercury, cadmium, arsenic, or other metals; poisoning by carbon monoxide, hydrogen sulfide or other gases; poisoning by benzol, carbon tetrachloride, or other organic solvents; poisoning by insecticide sprays such as parathion, lead arsenate; poisoning by other chemicals such as formaldehyde, plastics and resins; etc.

(25) Disorders Due to Physical Agents (Other Than Toxic Materials)

Examples: Heatstroke, sunstroke, heat exhaustion and other effects of environmental heat; freezing, frostbite and effects of exposure to low temperatures; caisson disease; effects of ionizing radiation (isotopes, X-rays, radium); effects of nonionizing radiation (welding flash, ultraviolet rays, microwaves, sunburn); etc.

(26) Disorders Associated With Repeated Trauma

Examples: Noise-induced hearing loss; synovitis, tenosynovitis, and bursitis; Raynaud's phenomena; and other conditions due to repeated motion, vibration or pressure.

(29) All Other Occupational Illnesses

Examples: Anthrax, brucellosis, infectious hepatitis, malignant and benign tumors, food poisoning, histoplasmosis, coccidioidomycosis, etc.

MEDICAL TREATMENT includes treatment (other than first aid) administered by a physician or by registered professional personnel under the standing orders of a physician. Medical treatment does NOT include first aid treatment (one-time treatment and subsequent observation of minor scratches, cuts, burns, splinters, and so forth, which do not ordinarily require medical care) even though provided by a physician or registered professional personnel.

ESTABLISHMENT: A single physical location where business is conducted or where services or industrial operations are performed (for example: a factory, mill, store, hotel, restaurant, movie theater, farm, ranch, bank, sales office, warehouse, or central administrative office). Where distinctly separate activities are performed at a single physical location (such as contract construction activities operated from the same physical location as a lumber yard), each activity shall be treated as a separate establishment.

For firms engaged in activities such as agriculture, construction, transportation, communications, and electric, gas and sanitary services, which may be physically dispersed, records may be maintained at a place to which employees report each day.

Records for personnel who do not primarily report or work at a single establishment, such as traveling salesmen, technicians, engineers, etc., shall be maintained at the location from which they are paid or the base from which personnel operate to carry out their activities.

WORK ENVIRONMENT is comprised of the physical location, equipment, materials processed or used, and the kinds of operations carried out by an employee in the performance of his work, whether on or off the employer's premises.

OSHA No. 101 Form approved
Case or File No. .... OMB No. 44R 1453

SUPPLEMENTARY RECORD OF OCCUPATIONAL INJURIES AND ILLNESSES

EMPLOYER

1. Name ....
2. Mail address ....
   (No. and street) (City or town) (State)
3. Location, if different from mail address ....

INJURED OR ILL EMPLOYEE

4. Name .... Social Security No. ....
   (First) (Middle) (Last)
5. Home address ....
   (No. and street) (City or town) (State)
6. Age .... 7. Sex: Male........ Female........ (Check one)
8. Occupation ....
   (Enter regular job title, *not* the specific activity he was performing at time of injury.)
9. Department ....
   (Enter name of department or division in which the injured person is regularly employed, even though he may have been temporarily working in another department at the time of injury.)

THE ACCIDENT OR EXPOSURE TO OCCUPATIONAL ILLNESS

10. Place of accident or exposure ....
    (No. and street)

    ....
    (City or town) (State)

    If accident or exposure occurred on employer's premises, give address of plant or establishment in which it occurred. Do not indicate department or division within the plant or establishment. If accident occurred outside employer's premises at an identifiable address, give that address. If it occurred on a public highway or at any other place which cannot be identified by number and street, please provide place references locating the place of injury as accurately as possible.
11. Was place of accident or exposure on employer's premises? .....
    ....
    (Yes or No)
12. What was the employee doing when injured? ....
    (Be specific. If he was
    ....
    using tools or equipment or handling material, name them and
    ....
    tell what he was doing with them.)

13. How did the accident occur? ....
(Describe fully the events which resulted
....
in the injury or occupational illness. Tell what happened and how it happened.
....
Name any objects or substances involved and tell how they were involved. Give full
....
details on all factors which led or contributed to the accident. Use separate sheet
....
for additional space.)

OCCUPATIONAL INJURY OR OCCUPATIONAL ILLNESS

14. Describe the injury or illness in detail and indicate the part of body affected ....
(e.g.: amputation of right index finger at second joint;
....
fracture of ribs; lead poisoning; dermatitis of left hand, etc.)

15. Name the object or substance which directly injured the employee. (For example, the machine or thing he struck against or which struck him; the vapor or poison he inhaled or swallowed; the chemical or radiation which irritated his skin; or in cases of strains, hernias, etc., the thing he was lifting, pulling, etc.) ....
....
....

16. Date of injury or initial diagnosis of occupational illness ....
....
(Date)

17. Did employee die? .... (Yes or No)

OTHER

18. Name and address of physician ....
....

19. If hospitalized, name and address of hospital ....
....

Date of report .... Prepared by ....
Official position ....

SUPPLEMENTARY RECORD OF OCCUPATIONAL INJURIES AND ILLNESSES

To supplement the Log of Occupational Injuries and Illnesses (OSHA No. 100), each establishment must maintain a record of each recordable occupational injury or illness. Workmen's compensation, insurance, or other reports are acceptable as records if they contain all facts listed below or are supplemented to do so. If no suitable report is made for other purposes, this form (OSHA No. 101) may be used or the necessary facts can be listed on a separate plain sheet of paper. These records must also be available in the establishment without delay and at reasonable times for examination by representatives of the Department of Labor and the Department of Health, Education and Welfare, and states accorded jurisdiction under the act. The records must be maintained for a period of not less than five years following the end of the calendar year to which they relate.

Such records must contain at least the following facts:

1) *About the employer*—name, mail address, and location if different from mail address.

2) *About the injured or ill employee*—name, social security number, home address, age, sex, occupation, and department.

3) *About the accident or exposure to occupational illness*—place of accident or exposure, whether it was on employer's premises, what the employee was doing when injured, and how the accident occurred.

4) *About the occupational injury or illness*—description of the injury or illness, including part of body affected; name of the object or substance which directly injured the employee; and date of injury or diagnosis of illness.

5) *Other*—name and address of physician; if hospitalized, name and address of hospital; date of report; and name and position of person preparing the report.

SEE *DEFINITIONS* ON THE BACK OF OSHA FORM 100.

OSHA No. 102

Complete no later than one month after close of calendar year. See back of this form for posting requirements and instructions.

Form Approved
OMB No. 44R 1453

SUMMARY
OF
OCCUPATIONAL INJURIES AND ILLNESSES
FOR CALENDAR YEAR 19__

Use previous edition of this form for summarizing your 1974 cases. This edition is for summarizing your cases for 1975 and subsequent years.

Establishment:
NAME ______________________
ADDRESS ______________________

INJURY AND ILLNESS CATEGORY			TOTAL CASES	DEATHS	LOST WORKDAY CASES				NONFATAL CASES WITHOUT LOST WORKDAYS	TERMINATIONS OR PERMANENT TRANSFERS
					Total Lost Workday Cases	Cases Involving Days Away From Work	Days Away From Work	Days of Restricted Work Activity		
	CATEGORY	CODE	Number of entries in Col. 7 of the log. (1)	Number of entries in Col. 8 of the log. (2)	Number of checks in Col. 9 of the log. (3)	Number of entries in Col. 9A of the log. (4)	Sum of entries in Col. 9A of the log. (5)	Sum of entries in Col. 9B of the log. (6)	Number of checks in Col. 10 of the log. (7)	Number of checks in Col. 11 of the log. (8)
OCCUPATIONAL INJURIES		10								
OCCUPATIONAL ILLNESSES	Occupational Skin Diseases or Disorders	21								
	Dust Diseases of the Lungs	22								
	Respiratory Conditions Due to Toxic Agents	23								
	Poisoning (Systemic Effects of Toxic Materials)	24								
	Disorders Due to Physical Agents	25								
	Disorders Associated With Repeated Trauma	26								
	All Other Occupational Illnesses	29								
	TOTAL—OCCUPATIONAL ILLNESSES (Sum of codes 21 through code 29)	30								
TOTAL—OCCUPATIONAL INJURIES AND ILLNESSES (Sum of code 10 and code 30)		31								

This is NOT a report form. Keep it in the establishment for 5 years.

I certify that this Summary of Occupational Injuries and Illnesses is true and complete, to the best of my knowledge.

Signature ______________________
Title ______________________
Date ______________________

Reduced Size Facsimile

SUMMARY OF OCCUPATIONAL INJURIES AND ILLNESSES

Every employer who is subject to the recordkeeping requirements of the Occupational Safety and Health Act of 1970 must use this form to prepare an annual summary of the occupational injury and illness experience of the employees in each of his establishments within one month following the end of each year.

POSTING REQUIREMENTS: A copy or copies of the summary must be posted at each establishment in the place or places where notices to employees are customarily posted. This summary must be posted no later than February 1 and must remain in place until March 1.

INSTRUCTIONS for completing this form: All entries must be summarized from the log (OSHA No. 100) or its equivalent. Before preparing this summary, review the log to be sure that entries are correct and each case is included in only one of the following classes: deaths (date in column 8), lost workday cases (check in column 9), or nonfatal cases without lost workdays (check in column 10). If an employee's loss of workdays is continuing at the time the summary is being made, estimate the number of future workdays he will lose and add that estimate to the workdays he has already lost and include this total in the summary. No further entries are to be made with respect to such cases in the next year's summary.

Occupational injuries and the seven categories of occupational illnesses are to be summarized separately. Identify each case by the code in column 7 of the log of occupational injuries and illnesses.

The summary from the log is made as follows:

A. For occupational injuries (identified by a code 10 in column 7 of the log form) make entries on the line for code 10 of this form.

Column 1—Total Cases. Count the number of entries which have a code 10 in column 7 of the log. Enter this total in column 1 of this form. This is the total of occupational injuries for the year.

Column 2—Deaths. Count the number of entries (date of death) for occupational injuries in column 8 of the log.

Column 3—Total Lost Workday Cases. Count the number of checks for occupational injuries in column 9 of the log.

Column 4—Cases Involving Days Away From Work. Count the number of entries for occupational injuries in column 9A of the log.

Column 5—Days Away From Work. Add the entries (total days away) for occupational injuries in column 9A of the log.

Column 6—Days of Restricted Work Activity. Add the entries (total of such days) for occupational injuries in column 9B of the log.

Column 7—Nonfatal Cases Without Lost Workdays. Count the number of checks for occupational injuries in column 10 of the log.

Column 8—Terminations or Permanent Transfers. Count the number of checks for occupational injuries in column 11 of the log.

CHECK: If the totals for code 10 have been entered correctly, the sum of columns 2, 3, and 7 will equal the number entered in column 1.

B. Follow the same procedure for each illness code, entering the totals on the appropriate line of this form.

C. Add the entries for codes 21 through 29 in each column for occupational illnesses and enter totals on the line for code 30.

D. Add the entries for codes 10 and 30 in each column and enter totals on the line for code 31.

CHECK: If the summary has been made correctly, the entry in column 1 of the total line (code 31) of this form will equal the total number of cases on the log.

The person responsible for the preparation of the summary shall certify that it is true and complete by signing the statement on the form.

Use previous edition of this form for summarizing your 1974 cases. This edition is for summarizing your cases for 1975 and subsequent years. Forms for the 1974 summary can be obtained from the appropriate State statistical grant agency (if there is one in your State) or from the appropriate Regional Office of the Bureau of Labor Statistics. Addresses are in the booklet entitled Recordkeeping Requirements under the Occupational Safety and Health Act of 1970.

OSHA No. 103
U.S. DEPARTMENT OF LABOR
Bureau of Labor Statistics
for the Occupational Safety
and Health Administration
Washington, D.C. 20212

THIS REPORT IS MANDATORY UNDER PUBLIC LAW 91-596
IT WILL BE USED ONLY FOR ADMINISTRATIVE AND STATISTICAL PURPOSES

OMB APPROVAL NO. 44-R1492
Approval Expires December 1974

St. Sch. # Ck. Suf. Cd. SIC Edit SIC Wt.

1973 OCCUPATIONAL INJURIES AND ILLNESSES SURVEY

(Covering Calendar Year 1973)

COMPLETE THIS REPORT WHETHER OR NOT THERE WERE ANY RECORDABLE OCCUPATIONAL INJURIES OR ILLNESSES. READ INSTRUCTIONS BEFORE COMPLETING THIS FORM

THIS IS YOUR FILE COPY DO NOT RETURN

I. ESTABLISHMENTS INCLUDED IN THIS REPORT
This report should include only those establishments located in, or identified by, the Report Location or Identification which appears below your mailing address on this form. Enter the number of establishments (see definition on page 1) included in this report: ______

II. AVERAGE EMPLOYMENT IN 1973
Enter the average number of employees during calendar year 1973. Count all classes of employees, including seasonal, temporary, part-time, etc. See instructions for examples of computing your average employment.
(Round to the nearest whole number) ______

III. TOTAL HOURS WORKED IN 1973
Enter the total number of hours actually worked by all employees during 1973. DO NOT include any non-work time even though paid, such as vacations, sick leave, holidays, etc.
(Round to the nearest whole number) ______

IV. SUPPORT ACTIVITIES PERFORMED FOR OTHER ESTABLISHMENTS OF YOUR COMPANY
Does this report include any establishment (s) whose primary function is to provide support activities or services exclusively for other establishments of your company?
(1) ☐ No (2) ☐ Yes
If yes, indicate the primary type of service or support provided (check as many as apply).
(1) ☐ Central administrative office
(2) ☐ Research, development, or testing
(3) ☐ Storage (warehouse)
(4) ☐ Other - Specify ______

V. NATURE OF BUSINESS FOR 1973
1. Indicate the general type of activity performed during 1973 by the establishment(s) included in this report (i.e., manufacturing, wholesale trade, retail trade, construction, services, finance, etc.):
______
______
______

2. Enter in order of importance the principal products manufactured, lines of trade, specific services, or other description of specific activities for 1973. — For each entry, also include the approximate percent of total 1973 annual value of production, sales, or receipts.

(1) ______ ___ %
(2) ______ ___ %
(3) ______ ___ %
(4) ______ ___ %
(5) ______ ___ %
(6) ______ ___ %

VI. RECORDABLE INJURIES AND ILLNESSES
Did you have any recordable injuries or illnesses during calendar year 1973? (Check one)
(1) ☐ No - complete Section VII, Part B and Section IX
(2) ☐ Yes - complete Sections VII, VIII and IX

VII. MONTHLY DATA OF RECORDABLE INJURIES AND ILLNESSES
A. Of the Total Recordable Occupational Injuries and Illnesses (Section VIII, Line 31 columns 3, 4, and 7), how many occurred in the following months?

Calendar Year 1973

Jan.	______	July	______
Feb.	______	Aug.	______
Mar.	______	Sept.	______
Apr.	______	Oct.	______
May	______	Nov.	______
June	______	Dec.	______

B. If your establishment(s) had an OSHA compliance inspection during calendar year 1973, please enter the month of inspection ______ . ______

REPORT LOCATION OR IDENTIFICATION ⟶

BUREAU OF LABOR STATISTICS
OCCUPATIONAL SAFETY AND HEALTH SURVEY
441 G STREET, NW.
WASHINGTON, D.C. 20212

FOR INFORMATION CALL: (202) 523-1221

Reduced Size Facsimile

VIII. INJURY AND ILLNESS SUMMARY (Covering Calendar Year 1973)

INSTRUCTIONS:

- This section may be completed by Copying data from OSHA Form No. 102 "Summary, Occupational Injuries and Illnesses" which you are required to complete and post in your establishment.
- Leave Section VIII blank if there were no recordable injuries or illnesses during 1973.
- Code 30 - Add all Occupational Illnesses (Code 21+ 22+ 23+ 24+ 25+ 26+ 29) and enter on this line for each column (3) through (8).
- Code 31 - Add Occupational Injuries (Code 10) and the sum of all Occupational Illnesses (Code 30) and enter on this line for each column (3) through (8).

Code (1)	Category (2)	FATALITIES (deaths) (3)	LOST WORKDAY CASES: Number of Cases (4)	LOST WORKDAY CASES: Number of Cases Involving Permanent Transfer to Another Job or Termination of Employment (5)	LOST WORKDAY CASES: Number of Lost Workdays (6)	NONFATAL CASES WITHOUT LOST WORKDAYS*: Number of Cases (7)	NONFATAL CASES WITHOUT LOST WORKDAYS*: Number of Cases Involving Transfer to Another Job or Termination of Employment (8)
10	OCCUPATIONAL INJURIES						
21	OCCUPATIONAL ILLNESSES: Occupational Skin Diseases or Disorders						
22	Dust Diseases of the Lungs (Pneumoconioses)						
23	Respiratory Conditions Due To Toxic Agents						
24	Poisoning (Systemic Effects of Toxic Materials)						
25	Disorders Due To Physical Agents (Other Than Toxic Materials)						
26	Disorders Due To Repeated Trauma						
29	All Other Occupational Illnesses						
30	SUM of ALL OCCUPATIONAL ILLNESSES (Add Codes 21 thru 29)						
31	TOTAL OF ALL OCCUPATIONAL INJURIES AND ILLNESSES (Add Codes 10 + 30)						

* Nonfatal Cases Without Lost Workdays - Cases resulting in: Medical treatment beyond first-aid, diagnosis of occupational illness, loss of consciousness, restriction of work or motion, or transfer to another job (without lost workdays).

COMMENTS: ______________________________

IX. Report Prepared By: ______________ Date: ______________

Title: ______________ Area Code and Phone: ______________

Reduced Size Facsimile

11

MANDATORY POSTING REQUIREMENTS

MANDATORY POSTING REQUIREMENTS

The law requires the posting of certain notices and information.

> • **NOTE** • Violation of posting requirements is subject to a penalty of $1,000 each maximum. See Chapter 5.

WHAT MUST BE POSTED?

Job poster: The "Safety and Health Protection on the Job" poster must be posted in a conspicuous place in each establishment or workplace where employees usually report to work. (See reduced-size facsimile at the end of this chapter.) *NOTE:* Copies of required full-size posters may be obtained at the nearest OSHA office. (See the Directory at the back of this manual.)

Annual summary: For requirements for posting the annual summary (OSHA Form 102), see Chapter 10.

Variance applications posted: For requirements for posting, variance applications, see Chapter 7 [Section 6.(b)(C)(B)(V)].

Citations and notices posted: For requirements for posting citations and notices, see Chapter 5 [Section 9.(b)].

Other notices posted: Other notices may be required to be posted by rule, standard, etc. Such notices shall be posted per instruction or directions.

PROVISION OF THE LAW REQUIRING POSTING

"AS STATED"
Sec. 8. "THE ACT"
Sec. 8.(c)(1) *Each employer shall make, keep and pre-*

safety and health protection on the job

The Williams-Steiger Occupational Safety and Health Act of 1970 provides job safety and health protection for workers through the promotion of safe and healthful working conditions throughout the Nation. Requirements of the Act include the following:

Employers: **Each employer shall furnish to each of his employees employment and a place of employment free from recognized hazards that are causing or are likely to cause death or serious harm to his employees; and shall comply with occupational safety and health standards issued under the Act.**

Employees: **Each employee shall comply with all occupational safety and health standards, rules, regulations and orders issued under the Act that apply to his own actions and conduct on the job.**

days, or until it is corrected, whichever is later, to warn employees of dangers that may exist there.

Proposed Penalty: The Act provides for mandatory penalties against employers of up to $1,000 for each serious violation and for optional penalties of up to $1,000 for each nonserious violation. Penalties of up to $1,000 per day may be proposed for failure to correct violations within the proposed time period. Also, any employer who willfully or repeatedly violates the Act may be assessed penalties of up to $10,000 for each such violation.

Criminal penalties are also provided for in the Act. Any willful violation resulting in death of an employee, upon conviction, is punishable by a fine of not more than $10,000 or by imprisonment for not more than six months, or by both. Conviction of an employer after a first conviction doubles these maximum penalties.

Reduced Size Facsimile

Washington, D. C.
1975
OSHA 2003

Secretary of Labor

U. S. Department of Labor
Occupational Safety and Health Administration

The Occupational Safety and Health Administration (OSHA) of the Department of Labor has the primary responsibility for administering the Act. OSHA issues occupational safety and health standards, and its Compliance Safety and Health Officers conduct jobsite inspections to ensure compliance with the Act.

Inspection:

The Act requires that a representative of the employer and a representative authorized by the employees be given an opportunity to accompany the OSHA inspector for the purpose of aiding the inspection.

Where there is no authorized employee representative, the OSHA Compliance Officer must consult with a reasonable number of employees concerning safety and health conditions in the workplace.

Complaint:

Employees or their representatives have the right to file a complaint with the nearest OSHA office requesting an inspection if they believe unsafe or unhealthful conditions exist in their workplace. OSHA will withhold names of employees complaining on request.

The Act provides that employees may not be discharged or discriminated against in any way for filing safety and health complaints or otherwise exercising their rights under the Act.

An employee who believes he has been discriminated against may file a complaint with the nearest OSHA office within 30 days of the alleged discrimination.

Citation:

If upon inspection OSHA believes an employer has violated the Act, a citation alleging such violations will be issued to the employer. Each citation will specify a time period within which the alleged violation must be corrected.

The OSHA citation must be prominently displayed at or near the place of alleged violation for three

While providing penalties for violations, the Act also encourages efforts by labor and management, before an OSHA inspection, to reduce injuries and illnesses arising out of employment.

Voluntary Activity:

The Department of Labor encourages employers and employees to reduce workplace hazards voluntarily and to develop and improve safety and health programs in all workplaces and industries.

Such cooperative action would initially focus on the identification and elimination of hazards that could cause death, injury, or illness to employees and supervisors. There are many public and private organizations that can provide information and assistance in this effort, if requested.

More Information:

Additional information and copies of the Act, specific OSHA safety and health standards, and other applicable regulations may be obtained from the nearest OSHA Regional Office in the following locations:

Atlanta, Georgia

Boston, Massachusetts

Chicago, Illinois

Dallas, Texas

Denver, Colorado

Kansas City, Missouri

New York, New York

Philadelphia, Pennsylvania

San Francisco, California

Seattle, Washington

Telephone numbers for these offices, and additional Area Office locations, are listed in the telephone directory under the United States Department of Labor in the United States Government listing.

Reduced Size Facsimile

serve, and make available to the Secretary or the Secretary of Health, Education, and Welfare, such records regarding his activities relating to this Act as the Secretary, in cooperation with the Secretary of Health, Education, and Welfare, may prescribe by regulation as necessary or appropriate for the enforcement of this Act or for developing information regarding the causes and prevention of occupational accidents and illnesses. In order to carry out the provisions of this paragraph such regulations may include provisions requiring employers to conduct periodic inspections. ***The Secretary shall also issue regulations requiring that employers, through posting of notices or other appropriate means, keep their employees informed of their protections and obligations under this Act, including the provisions of applicable standards.***

12

OSHA SAFETY AND HEALTH STANDARDS

Provisions for and a Helpful Guide to the Numerous OSHA Standards

OSHA SAFETY AND HEALTH STANDARDS

EMPLOYERS AND EMPLOYEES MUST COMPLY WITH OSHA STANDARDS

The area of OSHA Safety and Health Standards is without question the most involved. Thousands of OSHA Safety and Health Standards have already been adopted and new standards are frequently added. Obviously, all standards do not apply to every workplace.

HOW CAN YOU DETERMINE AND FIND THOSE STANDARDS WHICH APPLY TO YOUR WORK?

First, OSHA standards are contained in five volumes and are available from the Superintendent of Documents, U.S. Government Printing Office, Washington, D.C. 20402. (See order form at the end of this chapter.) Copies of this order form are available at any OSHA office. The five volumes are as follows:

Volume I—General Industry Standards

Part 1910—This volume contains the general industry standards, which apply to practically every workplace in the nation. (See index of subparts herein.)

Volume II—Maritime Standards

Part 1915—Ship Repairing
Part 1916—Shipbuilding
Part 1917—Shipbreaking
Part 1918—Longshoring
This volume contains the standards for the maritime and related industries. (See index of subparts herein.)

Volume III—Construction Standards

Part 1926—This volume contains standards for the construction industry. (See index of subparts herein.)

Volume IV—Other Regulations and Procedures

This volume contains regulations and procedures covering a variety of regulations. (See index herein.)

Volume V—Compliance Operations Manual

This volume contains rules and regulations used by OSHA in carrying out its programs. (See index herein.)

Second, you should review the following index of all volumes to determine which standards apply to your workplace.

Third, you should obtain a copy of the applicable standards that affect your workplace.

OSHA standards are being enforced daily, with stiff fines being assessed for noncompliance. The following is a chart showing OSHA inspections, citations, and penalties for January 1975:

OSHA INSPECTION FIGURES—JANUARY 1975

Industry	Manuf.	Const.	Marit.	Trans.	Retail	Whsl.	Serv.	Mine	F–I–R*	Agri.	Year to date	1974
Employees covered	852,517	83,902	143,298	68,711	42,305	16,986	26,274	7,435	1,505	2,933	1,245,866	673,812
Inspections	3,785	2,081	232	385	875	339	359	58	22	20	8,156	5,882
Accident	43	55	7	12	4	3	11	7	0	5	147	135
Complaint	343	102	22	75	30	21	32	1	6	5	637	428
General	2,352	1,675	178	236	667	238	256	48	14	8	5,672	5,319
Followup	1,047	249	25	62	174	77	60	2	2	2	1,700	No figures
In compliance (%)	18	35	54	23	12	19	22	33	21	50	24	22
Citations	2,742	1,482	152	290	706	242	262	57	18	11	5,962	3,870
Violations	17,184	4,041	635	1,404	3,409	1,259	1,395	242	96	55	29,720	21,531
Serious	111	167	15	7	9	4	12	9	0	1	335	175
Nonserious	16,960	3,816	593	1,397	3,397	1,246	1,379	233	96	54	29,171	21,316
Willful, repeated, imminent danger	113	58	27	0	3	9	4	0	0	0	214	40
Failure to abate	124	4	0	2	21	21	5	0	0	0	177	No figures
Proposed $ penalties	324,806	240,519	29,090	17,150	35,825	13,144	22,076	8,830	1,015	910	693,365	475,682
Contested cases	69	59	10	14	26	4	5	2	0	0	189	140

* Finance–insurance–real estate.

The following is an index of the five volumes of OSHA standards:

VOLUME I—GENERAL INDUSTRY STANDARDS

Part 1910

SUBPART

A	General
B	Adoption and Extension of Established Federal Standards
C	(Reserved)
D	Walking–Working Surfaces
E	Means of Egress
F	Powered Platforms, Manlifts, and Vehicle-Mounted Platforms
G	Occupational Health and Environmental Control
H	Hazardous Materials
I	Personal Protective Equipment
J	General Environmental Controls
K	Medical and First Aid
L	Fire Protection
M	Compressed Gas and Compressed Air Equipment
N	Materials Handling and Storage
O	Machinery and Machine Guarding
P	Hand and Portable Powered Tools and Other Hand-Held Equipment
Q	Welding, Cutting, and Brazing
R	Special Industries
S	Electrical

VOLUME II—MARITIME STANDARDS

Part 1915—Ship Repairing

SUBPART

A	General Provisions
B	Explosives and Other Dangerous Atmospheres

SUBPART

- C Surface Preparation and Preservation
- D Welding, Cutting, and Heating
- E Scaffolds, Ladders, and Other Working Surfaces
- F General Working Conditions
- G Gear and Equipment for Rigging and Materials Handling
- H Tools and Related Equipment
- I Personal Protective Equipment
- J Ship's Machinery and Piping Systems
- K Portable, Unfired Pressure Vessels, Drums, and Containers, Other Than Ship's Equipment
- L Electrical Machinery

Part 1916—Shipbuilding

SUBPART

- A General Provisions
- B Explosive and Other Dangerous Atmospheres
- C Surface Preparation and Preservation
- D Welding, Cutting, and Heating
- E Scaffolds, Ladders, and Other Working Surfaces
- F General Working Conditions
- G Gear and Equipment for Rigging and Materials Handling
- H Tools and Related Equipment
- I Personal Protective Equipment
- J Ship's Machinery and Piping Systems
- K Portable, Unfired Pressure Vessels, Drums, and Containers, Other Than Ship's Equipment
- L Electrical Machinery

Part 1917—Shipbreaking

SUBPART

- A General Provisions
- B Explosive and Other Dangerous Atmospheres
- C (Reserved)
- D Welding, Cutting, and Heating
- E Scaffolds, Ladders, and Other Working Surfaces
- F General Working Conditions

SUBPART

G	Gear and Equipment for Rigging and Materials Handling
H	Tools and Related Equipment
I	Personal Protective Equipment

Part 1918—Longshoring

SUBPART

A	General Provisions
B	Gangways and Gear Certification
C	Means of Access
D	Working Surfaces
E	Opening and Closing Hatches
F	Ship's Cargo Handling Gear
G	Cargo Handling Gear and Equipment Other Than Ship's Gear
H	Handling Cargo
I	General Walking Conditions
J	Personal Protective Equipment

VOLUME III—CONSTRUCTION STANDARDS

Part 1926

SUBPART

A	General
B	General Interpretations
C	General Safety and Health Provisions
D	Occupational Health and Environmental Controls
E	Personal Protective and Life Saving Equipment
F	Fire Protection and Prevention
G	Signs, Signals, and Barricades
H	Materials Handling, Storage Use, and Disposal
I	Tools—Hand and Power
J	Welding and Cutting
K	Electrical
L	Ladders and Scaffolding
M	Floors and Wall Openings, and Stairways
N	Cranes, Derricks, Hoists, Elevators, and Conveyors

SUBPART	
O	Motor Vehicles, Mechanized Equipment, and Marine Operations
P	Excavations, Trenching, and Shoring
Q	Concrete, Concrete Forms, and Shoring
R	Steel Erection
S	Tunnels and Shafts, Caissons, Cofferdams, and Compressed Air
T	Demolition
U	Blasting and Use of Explosives
V	Power Transmission and Distribution
W	Rollover Protective Structures; Overhead Protection
X	Effective Dates

VOLUME IV—OTHER REGULATIONS AND PROCEDURES

PART (29CFR)	
1901	Procedures for State Agreements
1902	State Plans for the Development and Enforcement of State Standards
1903	Inspections, Citations, and Proposed Penalties
1904	Recording and Reporting Occupational Injuries and Illnesses
1905	Rules of Practice for Variances, Limitations, Variations, Tolerances, and Exemptions Under the Williams–Steiger Occupational Safety and Health Act of 1970
1906	Administration Witnesses and Documents in Private Litigation
1911	Rules of Procedure for Promulgating, Modifying, or Revoking Occupational Safety or Health Standards
1912	Advisory Committees on Standards
1913	Disclosure of Information
1919	Gear Certification
1920	Procedures for Variations from Safety and Health Regulations Under Longshoremen's and Harbor Workers' Compensation Act
1921	Rules of Practice in Enforcement Proceedings Under Section 41 of the Longshoremen's and Harbor Workers' Compensation Act

PART (29CFR)

1922 Investigational Hearings Under Section 41 of the Longshoremen's and Harbor Workers' Compensation Act

1923 Safety and Health Provisions for Federal Agencies

1924 Safety Standards Applicable to Workshops and Rehabilitation Facilities Assisted by Grants

1925 Safety and Health Standards for Federal Service Contracts

1950 Development and Planning Grants for Occupational Safety and Health

1951 Grants for Implementing Approved State Plans

1975 Coverage of Employers Under the Williams–Steiger Occupational Safety and Health Act of 1970

2200 Occupational Safety and Health Review Commission, for Rules of Procedure

PART (42CFR)

National Institute for Occupational Safety and Health

(CHAPTER 1)

(NIOSH) Health Hazard Evaluations (Subchapter G—Part 85)

P.L. 91-596

Occupational Safety and Health Act of 1970

VOLUME V—COMPLIANCE OPERATIONS MANUAL

CHAPTER

I Purpose, Organization, and Use of Manual

II Organization and Functional Responsibilities

III Duties and Responsibilities of the National Office, Regional Administrator, Area Director, and Compliance Safety and Health Officer (CSHO)

IV Compliance Programing

V General Inspection Procedures

VI Complaints, Discrimination, Catastrophe, and/or Fatality Investigations and Other Special Situations

VII Construction, Maritime, and Agriculture

VIII Violations

IX Imminent Danger

CHAPTER
X Citations
XI Penalties
XII Field Reporting Procedures and Forms
XIII Industrial Hygiene and Occupational Health
XIV Standards Interpretations and Change Requests
XV Training
XVI Safety and Health Promotion and Publications
XVII State Jurisdiction and State Plans
XVIII Agreements with Other Federal Agencies
XIX Federal Agency Programs
XX Research and Related Activities
XXI Annotated Case Studies
XXII Review Commission
XXIII Variances
XXIV Disclosure and Publicity

PROVISIONS FOR STANDARDS

The following provisions are set forth in the law for the purpose of establishing and applying the standards.

Promulgation of Standards

Provisions in the law for adopting existing standards are as follows:

"AS STATED"
Sec. 6. "THE ACT"
Sec. 6.(a) *Without regard to Chapter 5 of Title 5, United States Code, or to the other subsections of this section, the Secretary shall, as soon as practicable during the period beginning with the effective date of this Act and ending two years after such date, by rule promulgate as an occupational safety or health standard any national consensus standard, and any established Federal standard, unless he determines that the promulgation of such a standard would*

not result in improved safety or health for specifically designated employees. In the event of conflict among any such standards, the Secretary shall promulgate the standard which assures the greatest protection of the safety or health of the affected employees.

Promulgate, Modify, Revoke Standards

Standards may be adopted, modified, or revoked.

"AS STATED"
Sec. 6. "THE ACT"
Sec. 6.(b) *The Secretary may by rule promulgate, modify, or revoke any occupational safety or health standard in the following manner:*

Advisory Committee Recommendations

Information and recommendations submitted by interested parties shall be given consideration.

"AS STATED"
Sec. 6. "THE ACT"
Sec. 6.(b)(1) *Whenever the Secretary, upon the basis of information submitted to him in writing by an interested person, a representative of any organization of employers or employees, a nationally recognized standards-producing organization, the Secretary of Health, Education, and Welfare, the National Institute for Occupational Safety and Health, or a State or political subdivision, or on the basis of information developed by the Secretary or otherwise available to him, determines that a rule should be promulgated in order to serve the objectives of this Act, the Secretary may request the recommendations of an advisory committee appointed under Section 7 of this Act. The Secretary shall provide such an advisory committee with any proposals of his own or of the Secretary of Health, Education, and Welfare, together with all pertinent factual information developed by the Secretary or the Secretary of*

Health, Education, and Welfare, or otherwise available, including the results of research, demonstrations, and experiments. An advisory committee shall submit to the Secretary its recommendations regarding the rule to be promulgated within ninety days from the date of its appointment or within such longer or shorter period as may be prescribed by the Secretary, but in no event for a period which is longer than two hundred and seventy days.

Publication in Federal Register and Provision for Interested Party Comments and Input

You have thirty days after publication in the Federal Register of a proposed rule adoption or change to submit written data or comments.

"AS STATED"
Sec. 6. "THE ACT"
Sec. 6.(b)(2) *The Secretary shall publish a proposed rule promulgating, modifying, or revoking an occupational safety or health standard in the Federal Register and shall afford interested persons a period of thirty days after publication to submit written data or comments. Where an advisory committee is appointed and the Secretary determines that a rule should be issued, he shall publish the proposed rule within sixty days after the submission of the advisory committee's recommendations or the expiration of the period prescribed by the Secretary for such submission.*

Objections to Proposed Standards

You may file an objection to the adoption or change of a proposed rule and receive a hearing on your objection.

"AS STATED"
Sec. 6. "THE ACT"
Sec. 6.(b)(3) *On or before the last day of the period provided for the submission of written data or comments*

under paragraph (2), any interested person may file with the Secretary written objections to the proposed rule, stating the grounds therefor and requesting a public hearing on such objections. Within thirty days after the last day for filing such objections, the Secretary shall publish in the Federal Register a notice specifying the occupational safety or health standard to which objections have been filed and a hearing requested, and specifying a time and place for such hearing.

Sec. 6.(b)(4) *Within sixty days after the expiration of the period provided for the submission of written data or comments under paragraph (2), or within sixty days after the completion of any hearing held under paragraph (3), the Secretary shall issue a rule promulgating, modifying, or revoking an occupational safety or health standard or make a determination that a rule should not be issued. Such a rule may contain a provision delaying its effective date for such period (not in excess of ninety days) as the Secretary determines may be necessary to insure that affected employers and employees will be informed of the existence of the standard and of its terms and that employers affected are given an opportunity to familiarize themselves and their employees with the existence of the requirements of the standard.*

Toxic Material Standards

Standards shall be provided for dealing with toxic materials or harmful physical agents. (These standards are included in the various subsections of the OSHA standards; see the standards index at the beginning of this chapter.)

"AS STATED"
Sec. 6. "THE ACT"
Sec. 6.(b)(5) *The Secretary, in promulgating standards dealing with toxic materials or harmful physical agents under this subsection, shall set the standard which most adequately assures, to the extent feasible, on the basis of*

the best available evidence, that no employee will suffer material impairment of health or functional capacity even if such employee has regular exposure to the hazard dealt with by such standard for the period of his working life. Development of standards under this subsection shall be based upon research, demonstrations, experiments, and such other information as may be appropriate. In addition to the attainment of the highest degree of health and safety protection for the employee, other considerations shall be the latest available scientific data in the field, the feasibility of the standards, and experience gained under this and other health and safety laws. Whenever practicable, the standard promulgated shall be expressed in terms of objective criteria and of the performance desired.

Standards for Warnings, Labels, and Protective Equipment

Standards for warnings, labels, and protective equipment shall be provided. (These standards are included in the various subsections of the OSHA standards; see the standards index at the beginning of this chapter.)

"AS STATED"
Sec. 6. "THE ACT"
Sec. 6.(b)(7) *Any standard promulgated under this subsection shall prescribe the use of labels or other appropriate forms of warning as are necessary to insure that employees are apprised of all hazards to which they are exposed, relevant symptoms and appropriate emergency treatment, and proper conditions and precautions of safe use or exposure. Where appropriate, such standard shall also prescribe suitable protective equipment and control or technological procedures to be used in connection with such hazards and shall provide for monitoring or measuring employee exposure at such locations and intervals, and in such manner as may be necessary for the protection of employees. In addition, where appropriate, any such standard shall prescribe the type and frequency of medical examinations or other tests which shall be made avail-*

able, by the employer or at his cost, to employees exposed to such hazards in order to most effectively determine whether the health of such employees is adversely affected by such exposure. In the event such medical examinations are in the nature of research, as determined by the Secretary of Health, Education, and Welfare, such examinations may be furnished at the expense of the Secretary of Health, Education, and Welfare. The results of such examinations or tests shall be furnished only to the Secretary or the Secretary of Health, Education, and Welfare, and, at the request of the employee, to his physician. The Secretary, in consultation with the Secretary of Health, Education, and Welfare, may by rule promulgated pursuant to Section 553 of Title 5, United States Code, make appropriate modifications in the foregoing requirements relating to the use of labels or other forms of warning, monitoring, or measuring, and medical examinations, as may be warranted by experience, information, or medical or technological developments acquired subsequent to the promulgation of the relevant standard.

Differences with National Consensus Standards

OSHA shall publish reasons for adoption where differences occur.

"AS STATED"
Sec. 6. "THE ACT"
Sec. 6.(b)(8) *Whenever a rule promulgated by the Secretary differs substantially from an existing national consensus standard, the Secretary shall, at the same time, publish in the Federal Register a statement of the reasons why the rule as adopted will better effectuate the purposes of this Act than the national consensus standard.*

Temporary Standards

Temporary standards can be immediately adopted.

"AS STATED"
Sec. 6. "THE ACT"
Sec. 6.(c)

(1) *The Secretary shall provide, without regard to the requirements of Chapter 5, Title 5, United States Code, for an emergency temporary standard to take immediate effect upon publication in the Federal Register if he determines (A) that employees are exposed to grave danger from exposure to substances or agents determined to be toxic or physically harmful or from new hazards, and (B) that such emergency standard is necessary to protect employees from such danger.*

(2) Such standard shall be effective until superseded by a standard promulgated in accordance with the procedures prescribed in paragraph (3) of this subsection.

(3) Upon publication of such standard in the Federal Register the Secretary shall commence a proceeding in accordance with Section 6(b) of this Act, and the standard as published shall also serve as a proposed rule for the proceeding. The Secretary shall promulgate a standard under this paragraph no later than six months after publication of the emergency standard as provided in paragraph (2) of this subsection.

Publication Requirements

OSHA must publish information.

"AS STATED"
Sec. 6. "THE ACT"
Sec. 6.(e) *Whenever the Secretary promulgates any standard, makes any rule, order, or decision, grants any exemption or extension of time, or compromises, mitigates, or settles any penalty assessed under this Act, he shall include a statement of the reasons for such action, which shall be published in the Federal Register.*

Petition for Judicial Review

Any person who may be adversely affected by a standard may at any time within sixty days after such standard has been adopted file for a judicial review with the U.S. Court of Appeals.

"AS STATED"
Sec. 6. "THE ACT"
Sec. 6.(f) *Any person who may be adversely affected by a standard issued under this section may at any time prior to the sixtieth day after such standard is promulgated file a petition challenging the validity of such standard with the United States court of appeals for the circuit wherein such person resides or has his principal place of business, for a judicial review of such standard. A copy of the petition shall be forthwith transmitted by the clerk of the court to the Secretary. The filing of such petition shall not, unless otherwise ordered by the court, operate as a stay of the standard. The determinations of the Secretary shall be conclusive if supported by substantial evidence in the record considered as a whole.*

Priority for Establishing Standards

A priority for establishing standards is provided for.

"AS STATED"
Sec. 6. "THE ACT"
Sec. 6.(g) *In determining the priority for establishing standards under this section, the Secretary shall give due regard to the urgency of the need for mandatory safety and health standards for particular industries, trades, crafts, occupations, businesses, workplaces, or work environments. The Secretary shall also give due regard to the recommendations of the Secretary of Health, Education, and Welfare regarding the need for mandatory standards in determining the priority for establishing such standards.*

OCCUPATIONAL SAFETY AND HEALTH SUBSCRIPTION SERVICE

STANDARDS, INTERPRETATIONS, REGULATIONS, AND PROCEDURES

UNITED STATES DEPARTMENT OF LABOR
Occupational Safety and Health Administration
Washington, D. C. 20210

SUBSCRIPTION ORDER FORM

If more than one volume is ordered on this form, please anticipate a slight delay in processing.

Order forms received for single volumes will be processed more rapidly. This order form may be reproduced.

PLEASE ENTER MY SUBSCRIPTION FOR:

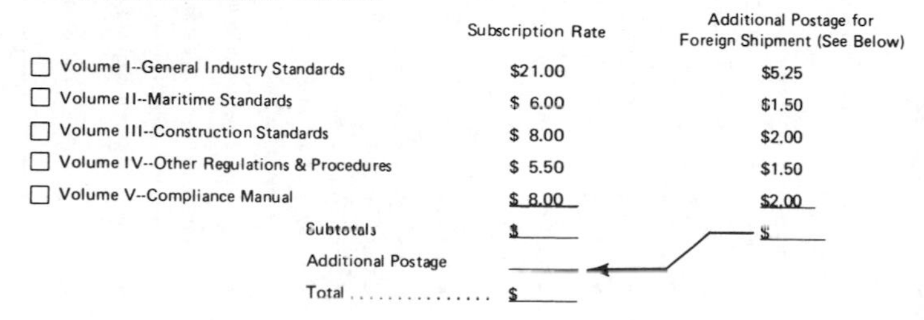

	Subscription Rate	Additional Postage for Foreign Shipment (See Below)
☐ Volume I--General Industry Standards	$21.00	$5.25
☐ Volume II--Maritime Standards	$ 6.00	$1.50
☐ Volume III--Construction Standards	$ 8.00	$2.00
☐ Volume IV--Other Regulations & Procedures	$ 5.50	$1.50
☐ Volume V--Compliance Manual	$ 8.00	$2.00
Subtotals	$ ____	$ ____
Additional Postage	____ ←	
Total	$ ____	

*No additional postage is required for mailing within the United States, its possessions, Canada, Mexico, and all Central and South American Countries except Argentina, Brazil, British Honduras, French Guiana, Guyana, and Surinam. For shipment to all other foreign countries include additional postage as quoted.

PLEASE PRINT

NAME – FIRST, LAST

COMPANY NAME OR ADDITIONAL ADDRESS LINE

STREET ADDRESS

CITY

STATE

ZIP CODE

☐ Remittance Enclosed (Make check payable to Superintendent of Documents)

☐ Charge to my Deposit Account No.________

MAIL ORDER FORM TO:
Superintendent of Documents,
Government Printing Office,
Washington, D.C. 20402

Reduced Size Facsimile

OCCUPATIONAL SAFETY AND HEALTH SUBSCRIPTION SERVICE STANDARDS, INTERPRETATIONS, REGULATIONS, AND PROCEDURES

The safety and health standards and regulations under the Williams-Steiger Occupational Safety and Health Act of 1970 are officially published in the Federal Register.

This subscription service supplements the Federal Register and the Code of Federal Regulations (29 CFR) by providing all of the standards, interpretations, regulations, and procedures in an easy-to-use, looseleaf form punched for a 3-ring binder.

Notices of changes and additions will be sent to subscribers to keep the subscription service current.

The subscription service is set in larger type and overall format of the standards is improved. Section numbers appear on the top of each page.

Volume I	General Industry Standards and Interpretations	$21.00
Volume II	Maritime Standards and Interpretations	$ 6.00
Volume III	Construction Standards and Interpretations	$ 8.00
Volume IV	Other Regulations and Procedures	$ 5.50
Volume V	Compliance Operations Manual	$ 8.00

Volume I--General Industry Standards (Part 1910)

- Fire protection
- Compressed gas and compressed air equipment
- Materials handling and storage
- Machinery and machine guarding
- Hand and portable powered tools and other hand-held equipment
- Welding, cutting, and brazing
- Powered platforms, manlifts, and vehicle-mounted platforms
- Occupational health and environmental control
- Hazardous materials
- Personal protective equipment
- Electrical code

Volume II--Maritime Standards

- Ship Repairing (Part 1915)
- Shipbuilding (Part 1916)
- Shipbreaking (Part 1917)
- Longshoring (Part 1918)

Volume III--Construction Standards (Part 1926)

Volume IV--Other Regulations and Procedures

- Procedures for State agreements
- State plans for development and enforcement of State standards
- Inspections, citations, and proposed penalties
- Recording and reporting occupational injuries and illnesses
- Rules of practice for variances, limitations variations, tolerances, and exemptions
- Rules of procedure for promulgating, modifying, or revoking occupational safety and health standards
- Advisory committees on standards
- Safety and health provisions for Federal agencies
- Safety standards applicable to workshops and rehabilitation facilities assisted by grants
- Safety and health standards for Federal service contracts
- Development and planning grants for occupational safety and health
- Occupational Safety and Health Review Commission
- NIOSH regulations

Volume V--Compliance Operations Manual

- Inspections
- Violations
- Penalties
- Field Reporting
- Citations

Reduced Size Facsimile

13

FEDERAL AGENCY PROGRAMS

FEDERAL AGENCY PROGRAMS

Government employers and employees must comply

Federal government employers and employees must comply with OSHA.

By Executive Order, all federal departments and agencies have been ordered to set up safety and health programs for their employees. Besides setting requirements for federal programs, the order calls for a Federal Advisory Council on Occupational Safety and Health to be established by the secretary of labor.

> • **NOTE** • Some approved state OSHA programs in effect require state and local government employers and employees to comply with their own state OSHA program, in which cases the federal OSHA programs do not apply.

OSHA programs for federal government employers and employees are covered under Section 19 of the act.

The provisions are as follows:

RESPONSIBILITY

Federal agencies must maintain effective occupational safety and health programs.

"AS STATED"
Sec. 19. "THE ACT"
Sec. 19.(a) *It shall be the responsibility of the head of each Federal agency to establish and maintain an effective and comprehensive occupational safety and health program which is consistent with the standards promulgated under Section 6. The head of each agency shall (after consultation with representatives of the employees thereof)—*

Provide Safe and Healthful Employment

Federal agencies must provide safe and healthful employment.

> "AS STATED"
> Sec. 19. "THE ACT"
> Sec. 19.(a)(1) *provide safe and healthful places and conditions of employment, consistent with the standards set under Section 6;*

Use Safety Equipment

Federal agencies must require the use of safety equipment.

> "AS STATED"
> Sec. 19. "THE ACT"
> Sec. 19.(a)(2) *acquire, maintain, and require the use of safety equipment, personal protective equipment, and devices reasonably necessary to protect employees;*

Maintain Records

Federal agencies must maintain records of accidents and illnesses. (See recordkeeping and reporting guidelines for federal agencies at the end of this chapter.)

> "AS STATED"
> Sec. 19. "THE ACT"
> Sec. 19.(a)(3) *keep adequate records of all occupational accidents and illnesses for proper evaluation and necessary corrective action;*

Establish Forms

Federal agencies must establish forms for recordkeeping. (See recordkeeping and reporting guidelines for federal agencies at the end of this chapter.)

"AS STATED"
Sec. 19. "THE ACT"
Sec. 19.(a)(4) *consult with the Secretary with regard to the adequacy as to form and content of records kept pursuant to subsection (a) (3) of this section; and*

Make Annual Report

Federal agencies must make an annual report of accidents and injuries. (See recordkeeping and reporting guidelines for federal agencies at the end of this chapter.)

"AS STATED"
Sec. 19. "THE ACT"
Sec. 19.(a)(5) *make an annual report to the Secretary with respect to occupational accidents and injuries and the agency's program under this section. Such report shall include any report submitted under Section 7902 (e) (2) of Title 5, United States Code.*

REPORT TO PRESIDENT AND CONGRESS

The secretary of labor shall report to the president on federal agencies' programs. (See recordkeeping and reporting guidelines for federal agencies at the end of this chapter.)

"AS STATED"
Sec. 19. "THE ACT"
Sec. 19.(b) *The Secretary shall report to the President a summary or digest of reports submitted to him under subsection (a) (5) of this section, together with his evaluations of and recommendations derived from such reports. The President shall transmit annually to the Senate and the House of Representatives a report of the activities of Federal agencies under this section.*

TITLE 5, U.S. CODE AMENDED

"AS STATED"
Sec. 19. "THE ACT"
Sec. 19.(c) *Section 7902 (c) (1) of Title 5, United States Code is amended by inserting after "agencies" the following: "and of labor organizations representing employees."*

ACCESS TO RECORDS

The secretary of labor shall have access to federal agencies' records.

"AS STATED"
Sec. 19. "THE ACT"
Sec. 19.(d) *The Secretary shall have access to records and reports kept and filed by Federal agencies pursuant to subsections (a)(3) and (5) of this section unless those records and reports are specifically required by Executive order to be kept secret in the interest of the national defense or foreign policy, in which case the Secretary shall have access to such information as will not jeopardize national defense or foreign policy.*

SAFETY AND HEALTH PROVISIONS FOR FEDERAL EMPLOYEES

PART 1960

Federal Register
Volume 39 Number 197
PART IV
WEDNESDAY, OCTOBER 9, 1974

DEPARTMENT OF LABOR

OCCUPATIONAL SAFETY AND HEALTH ADMINISTRATION
WASHINGTON, D.C.

Office of Federal Agency Safety Programs

Reduced Size Facsimile

EXECUTIVE ORDER

By an Executive Order of the President, all Federal employers and employees must comply with OSHA.

> • **NOTE** • Initially Executive Order No. 11612 dated July 26, 1971 was issued ordering Federal agency compliance. On September 30, 1974, Executive Order No. 11807 was issued which supersedes Executive Order No. 11612.

The following is Executive Order No. 11807.

EXECUTIVE ORDER NO. 11807
SEPTEMBER 30, 1974
OCCUPATIONAL SAFETY AND HEALTH
PROGRAMS FOR FEDERAL EMPLOYEES

"AS STATED"
As the Nation's largest employer, the Federal Government has a special obligation to set an example for all employers by providing a safe and healthful working environment for its employees.

For more than three years, the Federal Government has been seeking to carry out these solemn responsibilities under the terms of Executive Order No. 11612, issued in 1971 and based upon the authorities granted by the landmark Occupational Safety and Health Act of 1970 as well as section 7902(c) of title 5, United States Code.

Considerable progress has been achieved under the 1971 executive order, but it is now clear that even greater efforts are needed. It is therefore necessary that a new order be issued, reflecting this Nation's firm and renewed commitment to provide exemplary working conditions for those devoted to public service.

The provisions of this order are intended to ensure that each agency head is provided with all the guidance necessary to carry out an effective occupational safety and health program within the agency. Further, to keep the President abreast of progress, this order provides for detailed evaluations of the agencies' occupational safety and health programs by the Secretary of Labor and transmittal of those evaluations, together with agency comments, to the President. In addition, the Federal Safety Advisory Council on Occupational Safety and Health is continued because of its demonstrated value as an advisory body to the Secretary of Labor.

Experience has shown that agency heads desire and need more detailed guidance from the Secretary of Labor to make their occupational safety and health programs more effective. This order provides that the Secretary of Labor shall issue detailed guidelines and provide such further assistance as the agencies may request.

NOW, THEREFORE, by virtue of the authority vested in me by section 7902 (c)(1) of title 5 of the United States Code, and as President of the United States, it is hereby ordered as follows:

SCOPE OF THIS ORDER

This order applies to the Executive Branch of the Federal Government as well as *any* employing unit of the Federal Government.

"AS STATED"
Sec. 1. "THE ORDER"
Sec. 1. *For the purposes of this order, the term "agency" means an Executive Department, as defined in 5 U.S.C. 101, or any employing unit or authority of the Government of the United States not within an Executive Department. This order applies to all agencies of the Executive Branch of the Government; and by agreement between the Secretary of Labor (hereinafter referred to as the Secretary)*

and the head of an agency of the Legislative or Judicial Branches of the Government, the provisions of this order may be made applicable to such agencies. In addition, by agreement between the Secretary of Labor and the head of any agency, and to the extent permitted by law, the provisions of this order may be extended to employees of agencies who are employed in geographic locations to which the Occupational Safety and Health Act of 1970 is not applicable.

Duties of Heads of Agencies:

The head of each Federal agency shall:

- Establish and maintain safety and health program
- Designate or appoint an agency official to be responsible for agency programs
- Establish management information systems
- Establish procedures for adoption of and compliance with safety and health standards
- Provide for training
- Submit reports
- Cooperate with and assist the Secretary of Labor
- Observe guidelines

"AS STATED"
Sec. 2. "THE ORDER"
Sec. 2. *The head of each agency shall, after consultation with representatives of the employees thereof, establish and maintain an occupational safety and health program meeting the requirements of section 19 of the Occupational Safety and Health Act (hereinafter referred to as the act). In order to ensure that agency programs are consistent with the standards prescribed by section 6 of the act, the head of each agency shall:*

(1) Designate or appoint, to be responsible for the man-

agement and administration of the agency occupational safety and health program, an agency official with sufficient authority to represent effectively the interest and support of the agency head.

(2) Establish an occupational safety and health management information system, which shall include the maintenance of such records of occupational accidents, injuries, illnesses and their causes, and the compilation and transmittal of such reports based upon this information as the Secretary may require pursuant to section 3 of this order.

(3) Establish procedures for the adoption of agency occupational safety and health standards consistent with the standards promulgated by the Secretary pursuant to section 6 of the act; assure prompt attention to reports by employees or others of unsafe or unhealthful working conditions; assure periodic inspections of agency workplaces by personnel with sufficient technical competence to recognize unsafe and unhealthful working conditions in such workplaces; and assure prompt abatement of unsafe or unhealthful working conditions; including those involving facilities and/or equipment furnished by another Government agency, informing the Secretary of significant difficulties encountered in this regard.

(4) Provide adequate safety and health training for officials at the different management levels, including supervisory employees, employees responsible for conducting occupational safety and health inspections, and other employees. Such training shall include dissemination of information concerning the operation of the agency occupational safety and health program and the means by which each such person may participate and assist in the operation of that program.

(5) Submit to the Secretary on an annual basis a report containing such information as the Secretary shall prescribe.

(6) Cooperate with and assist the Secretary of Labor in the performance of his duties under section 19 of the act and section 3 of this order.

(7) Observe the guidelines published by the Secretary pursuant to section 3 of this order, giving due consideration to the mission, size and organization of the agency.

Duties of the Secretary of Labor:

The Secretary of Labor shall:

- Provide leadership and guidance
- Issue guidelines
- Prescribe recordkeeping and reporting requirements
- Provide consultation
- Perform for agencies
- Evaluate programs
- Submit reports

"AS STATED"
Sec. 3. "THE ORDER"
Sec. 3. *The Secretary shall provide leadership and guidance to the heads of agencies to assist them in fulfilling their occupational safety and health responsibilities by, among other means, taking the following actions:*

(1) Issue detailed guidelines to assist agencies in establishing and operating effective occupational safety and health programs appropriate to their individual missions, sizes, and organizations. Such guidelines shall reflect the requirement of section 19 of the Act for consultation with employee representatives.

(2) Prescribe recordkeeping and reporting requirements to enable agencies to assist the Secretary in meeting the requirements imposed upon him by section 24 of the act.

(3) Provide such consultation to agencies as the Secretary deems necessary and appropriate to ensure that agency standards adopted pursuant to section 2 of this order are consistent with the safety and health standards adopted by the Secretary pursuant to section 6 of the act; provide

leadership and guidance to agencies in the adequate occupational safety and health training of agency personnel; and facilitate the exchange of ideas and information throughout the Government with respect to matters of occupational safety and health through such arrangements as the Secretary deems appropriate.

(4) Perform for agencies, where deemed necessary and appropriate, the following services, upon request and reimbursement for the expenses thereof: (a) evaluate agency working conditions, and recommend to the agency head appropriate standards to be adopted pursuant to section 2 of this order to ensure that such working conditions are safe and healthful; (b) conduct inspections to identify unsafe or unhealthful working conditions, and provide assistance to correct such conditions; (c) train appropriate agency safety and health personnel.

(5) Evaluate the occupational safety and health programs of agencies, and submit to the President reports of such evaluations, together with agency responses thereto. These evaluations shall be conducted at least once annually for agencies employing more than 1,000 persons within the geographic locations to which the act applies, and as the Secretary deems appropriate for all other agencies, through such headquarters or field reviews as the Secretary deems necessary.

(6) Submit to the President each year a summary report of the status of the Federal agency occupational safety and health program, as well as analyses of individual agency progress and problems in correcting unsafe and unhealthful working conditions, together with recommendations for improving their performance.

FEDERAL ADVISORY COUNCIL ON OCCUPATIONAL SAFETY AND HEALTH

The Federal Advisory Council on Occupational Safety and Health is continued.

"AS STATED"
Sec. 4. "THE ORDER"
Sec. 4. (a) *The Federal Advisory Council on Occupational Safety and Health, established pursuant to Executive Order No. 11612, is hereby continued. It shall advise the Secretary in carrying out responsibilities under this order. This Council shall consist of fifteen members appointed by the Secretary and shall include representatives of Federal agencies and of labor organizations representing employees. At least five members shall be representatives of such labor organizations. The members shall serve for three-year terms with the terms of five members expiring each year, provided that this Council is renewed every two years in accordance with the Federal Advisory Committee Act. The members of the Federal Advisory Council on Occupational Safety and Health established pursuant to Executive Order No. 11612 shall be deemed to be its initial members under this order, and their terms shall expire in accordance with the terms of their appointments.*

(b) The Secretary, or a designee, shall serve as the Chairman of the Council, and shall prescribe such rules for the conduct of its business as he deems necessary and appropriate.

(c) The Secretary shall make available necessary office space and furnish the Council necessary equipment, supplies, and staff services, and shall perform such functions with respect to the Council as may be required by the Federal Advisory Committee Act.

Effect on Other Powers and Duties

Provisions of other effective orders, powers and duties are not interfered with.

"AS STATED"
Sec. 5. "THE ORDER"
Sec. 5. *Nothing in this order shall be construed to impair or alter the powers and duties of the Secretary or the heads*

of other Federal agencies pursuant to section 19 of the Occupational Safety and Health Act of 1970, sections 7901, 7902, and 7903 of title 5 of the United States Code, or any other provision of law, nor shall it be construed to alter the provisions of Executive Order No. 11491, as amended, Executive Order No. 11636, or other provisions of law providing for collective bargaining agreements and procedures. Matters of official leave for employee representatives involved in activities pursuant to this order shall be determined between each agency and these representatives pursuant to the procedures under Executive Order No. 11491, as amended, Executive Order No. 11636, or applicable collective bargaining agreements.

Termination of Existing Order

Executive Order No. 11612 is superseded.

"AS STATED"
Sec. 6. "THE ORDER"
Sec. 6. *Executive Order No. 11612 of July 26, 1971, is hereby superseded.*

SIGNED
Gerald R. Ford
THE WHITE HOUSE
September 28, 1974

SAFETY AND HEALTH PROVISIONS FOR FEDERAL EMPLOYEES, PART 1960, SUMMARY

Safety and Health Provisions for Federal Employees

Subpart A—General

1960.1	Purpose and Scope of this part.
1960.2	Definitions.

Subpart B—Recordkeeping and Reporting Requirements for Occupational Injuries, Illnesses and Accidents

1960.3	Purpose, scope and general provisions.
1960.4	Record or log of Federal occupational injuries and illnesses.
1960.5	Supplementary record of Federal occupational injuries and illnesses.
1960.6	Quarterly and annual summaries of Federal occupational injuries and illnesses.
1960.7	Quarterly and annual summaries of Federal occupational accidents.
1960.8	Reporting of serious accidents.
1960.9	Location and utilization of records and reports.
1960.10	Access to records by Secretary of Labor.
1960.11	Retention of records.
1960.12	Plan of action.
1960.13	[Reserved.]
1960.14	[Reserved.]

Subpart C—Agency Organization

1960.15	Purpose and scope.
1960.16	Designated safety and health official.
1960.17	Safety and health committees.
1960.18	Posting of notice; availability of Act, this part, and details of the agency safety and health program.
1960.19	Duties of agency officials and employees.
1960.20–24	[Reserved.]

Subpart D—Procedures for Inspections and Abatement

1960.25	Purpose, scope and general provisions.
1960.26	Safety and health inspectors; frequency of inspection.
1960.27	Conduct of inspection.
1960.28	Advance notice of inspections.
1960.29	Representatives of officials in charge and representatives of employees.
1960.30	Consultation with employees.

1960.31 Reports by employees of unsafe or unhealthful working conditions.
1960.32 Imminent danger.
1960.33 Notices of unsafe or unhealthful working conditions.
1960.34 Correction of unsafe or unhealthful working conditions.
1960.35–39 [Reserved.]

Subpart E—Agency Occupational Safety and Health Standards

1960.40 Purpose and scope.
1960.41 Procedures for adoption.
1960.42 Initial adoption of agency standards.
1960.43 Adoption of different and/or supplementary agency standards.
1960.44 Conflicting standards.
1960.45 Emergency standards.
1960.46 Access to standards.
1960.47–49 [Reserved.]

Subpart F—Field Federal Safety and Health Councils

1960.50 Purpose and scope.
1960.51–59 [Reserved.]

AUTHORITY: Sections 19 and 24 of the Occupational Safety and Health Act of 1970, 84 Stat. 1609, 1614, 29 U.S.C. 668, 673 and the provisions of Executive Order 11807.

SOURCE: The provisions of this Part 1960 appear in the *Federal Register*, Volume 39, Number 197, Part IV, Wednesday, October 9, 1974, issued at Washington, D.C., by the Office of Federal Agency Safety Programs, Occupational Safety and Health Administration, U.S. Department of Labor.

SUBPART A—GENERAL

1960.1 PURPOSE AND SCOPE OF THIS PART

(a) The primary purpose of the Occupational Safety and Health Act of 1970 is to assure safe and healthful working conditions for all employees in the Nation. While the enforcement procedures in sections 8, 9 and related sections of the Act do not apply to the Federal Government as an employer, Section 19 of the Act contains special provisions to afford protection to Federal employees. Under that section, it is the responsibility of the head of each Federal agency to establish and maintain an effective and comprehensive occupational safety and health program which is consistent with the standard promulgated under section 6 of the Act. The Secretary of Labor has important responsibilities in connection with the Federal agency occupational safety and health program, stemming from his duty under Section 19 to report to the President his evaluations and recommendations with respect to the programs of the various agencies. In addition, under Section 24 of the Act, the Secretary is directed to develop and maintain an effective program of collection, compilation and analysis of occupational safety and health statistics. To carry out that mandate, private-sector employers are required to file reports with the Secretary. While these reporting requirements do not apply to Federal agencies, the duties which section 24 imposes upon the Secretary of Labor necessarily extend to the collection, compilation and analysis of occupational safety and health statistics from the Federal Government, so that the Secretary may carry out the mandate of section 24 to conduct a comprehensive statistical program of job related injuries.

(b) The earlier Executive Order, No. 11612, issued on July 26, 1971 to implement the provisions of section 19 of the Act, has been reconsidered in light of experience, and a new Executive Order 11807 was issued on September 28, 1974 to replace it. Under the new Executive Order, certain detailed responsibilities of the heads of agencies are set forth, and the Secretary of Labor is required to issue recordkeeping and reporting regulations to carry out the provisions of Section 24 of the Act. In addition, the Secretary is required to issue guidelines which the heads of agencies are required to observe, taking into account the mission, size and organization of the agency. The purpose of this part is to carry out the requirements that the

Secretary issue regulations and guidelines for the safety and health programs of the various federal agencies.

(c) Since, under section 24 of the Act and section 3(2) of the Order, the Secretary is authorized to prescribe requirements for the agencies with respect to recordkeeping and reporting, the provisions in Subpart B of this part have generally been phrased in mandatory terms. The remaining subparts are guidelines to the agencies to assist them establish and operate effective safety and health programs. While the guidelines are phrased in nonmandatory terms, it should be emphasized that under terms of the new Executive Order the heads of the agencies are required to observe the guidelines, taking into consideration the mission, size and organization of the agency. It is the view of the Secretary that these guidelines will constitute a framework for a strong occupational safety and health program for Federal employees.

(d) Under the new Executive Order the Secretary is required to perform various services for the agencies. Agencies are encouraged to seek the assistance of the Secretary as needed to comply with the guidelines of this part and to otherwise operate effective safety and health programs for their employees. In addition, the Secretary will seek, with the cooperation of agency heads, to establish permanent channels between the Office of Federal Agency Safety Programs, Occupational Safety and Health Administration, U.S. Department of Labor, and Federal agencies to enable the purposes of this part to be effectuated, including agreements concerning the transmittal by agencies of information needed by the Secretary as set forth herein. Upon the request of an agency, the Office of Federal Agency Safety Programs will review proposed agency plans for the implementation of the provisions of this part in order to assure that such plans are in conformity with the intent of these provisions. To further aid the implementation of these guidelines, the Department of Labor will take steps to prepare and distribute handbooks to assist agencies in observing the guidelines of the Secretary in accordance with individual agency missions, sizes and organizations.

(e) The new Executive Order requires the heads of agencies to consult with representatives of employees and to provide for employee participation in the operation of agency safety and health programs. This requirement reflects the provisions of section 19 of the Act. In order to provide agencies with some guidance in this regard, the provisions of this part suggest specific instances where participation by employees and their representatives is particularly important in the operation of an agency's safety and health program. Such participation by employees and their representatives is separate but consistent with the provisions of other Executive Orders dealing with labor-management relations within the Federal Government.

(f) The regulations and guidelines of this part are applicable only to Federal employees and do not apply to employees of private contractors performing work under Government contracts, regardless of whether such privately employed workers perform their duties in Government-owned or-leased facilities, with government equipment, and together with government personnel. Protection of employees of private contractors is assured under the other provisions of the Act. Although this part does not make provision for the inclusion of Federal contractors nor their employees in agency safety and health programs, except as provided in § 1960.8 for reporting of serious accidents, safety and health programs operated pursuant to this part will offer some incidental protection to contractor employees working with Federal employees. Some agencies may wish to make further arrangements with such contractors to promote the safety and health of contractor employees when they are engaged in joint operations with Federal personnel. Agencies who wish to make such arrangements would be well advised to consult with their legal and budgetary personnel in this regard. Further, no such arrangement shall operate to relieve Federal contractors or their employees of any rights or responsibilities under the provisions of the Act, including compliance activities conducted by the Department of Labor or other appropriate authority.

1960.2 DEFINITIONS

(a) "Act" means the Williams-Steiger Occupational Safety and Health Act of 1970 (Stat. 1590 et seq., 29 U.S.C. 651 et seq).

(b) The term "agency" for the purposes of this part means an Executive Department, as defined in 5 U.S.C. 101, or any employing unit or authority of the government of the United States not within an Executive Department to which the provisions of Executive Order 11807 are applicable.

(c) The term "employee" as used in this part means any person employed or otherwise suffered, permitted or required to work by an "agency" as the latter term is defined in paragraph (b) of this section including non-civilian personnel.

(d) As used in Executive Order 11807, the term "consultation with representatives of the employees thereof" shall include such consultation, conference, or negotiation with representatives of agency employees as is consistent with Executive Order 11491, as amended, Executive Order 11636, or other collective bargaining arrangement. As used in this part, the term "representative of employees" should be interpreted with due regard for any obligation imposed by the aforementioned Executive Orders and any labor agreement that may cover the employees involved.

(e) The term "establishment" means a single physical location where business is conducted or where services or operations are performed. Where distinctly separate activities are performed at a single physical location, each activity shall be treated as a separate establishment.

(f) The term "reporting unit" means an establishment, except as otherwise agreed between the agency and the Office of Federal Agency Safety Programs, U.S. Department of Labor. Any such agreement in effect prior to the promulgation of this part shall remain in effect unless either party desires modification.

(g) The term "designated safety and health official" means the individual who is responsible for the management of the safety and health program within his agency and is designated or appointed by the head of the agency pursuant to § 1960.16 and the provisions of Executive Order 11807.

(h) The term "safety and health specialist" means a person or persons who meet the Civil Service standards for Safety Manager/Specialist GS-018, Safety Engineer, GS-803, Fire Protection Engineer GS-804, Industrial Hygienist GS-690, Fire Protection Specialist/Marshal GS-081, Health Physicist GS-1306, or equally qualified military, agency or nongovernment personnel.

(**i**) The term "safety and health inspector" means a safety and health specialist or other person authorized pursuant to § 1960.26 of this part to carry out inspections for the purpose of Subpart D of this part.

(**j**) The term "working days" means Mondays through Fridays (excluding Federal holidays), or other appropriate authorized days of agency operation.

(**k**) "Recordable occupational injuries or illnesses" are any occupational injuries or illnesses which result in:

1: Fatalities, regardless of the time between the injury and death, or the length of the illness; or

2: Cases, other than fatalities, that result in lost workdays; or

3: Non-fatal cases without lost workdays which result in transfer to another job or termination of employment, or require medical treatment (other than first aid), or involve loss of consciousness or restriction of work or motion. This category also includes any diagnosed occupational illnesses which are reported to the employer but are not classified as fatalities or lost workday cases.

(**l**) "Medical treatment" includes treatment administered by a physician or by registered professional personnel under the standing orders of a physician. Medical treatment does not include first aid treatment even though provided by a physician or registered professional personnel.

(**m**) "First aid" is any one-time treatment, and any followup visit for the purpose of observation, of minor scratches, cuts, burns, splinters, and so forth, which do not ordinarily require medical care. Such one-time treatment, and followup visit for the purpose of observation, is considered first aid even though provided by a physician or registered professional personnel.

(**n**) The term "lost workdays" means the number of days the employee would have worked but could not because of occupational injury or illness. The number of lost workdays should not include the day of injury. The number of days includes all days (consecutive or not) on which, because of the injury or illness:

1: the employee would have worked but could not, or
2: the employee was assigned to a temporary job, or

3: the employee worked at a permanent job less than full time, or

4: the employee worked at a permanently assigned job but could not perform all duties normally assigned to it.

For employees not having a regularly scheduled shift, i.e., certain truck drivers, construction workers, part-time employees, etc., it may be necessary to estimate the number of lost workdays. Estimates of lost workdays shall be based on prior work history of the employee and days worked by employees, not ill or injured, working in the agency and/or occupation of the ill or injured employee.

SUBPART B—RECORDKEEPING AND REPORTING REQUIREMENTS FOR OCCUPATIONAL INJURIES, ILLNESSES AND ACCIDENTS

1960.3 Purpose, Scope and General Provisions

(**a**) The purpose of this subpart is to establish uniform requirements for the collection and compilation by agencies of occupational safety and health data, thereby assisting the Secretary of Labor in meeting the requirement imposed upon him by Section 24 of the Act to "develop and maintain an effective program of collection, compilation, and analysis of occupational safety and health statistics," and enabling agencies to establish occupational safety and health management information systems pursuant to the requirements of Executive Order 11807.

(**b**) In order to perform his duties under Section 19 of the Act and Executive Order 11807, particularly with respect to providing the President with current information about the Federal agency safety and health program, it is necessary that the Secretary be promptly informed of serious accidents involving agency employees as provided in § 1960.8. Assistance to agencies in the investigation of such accidents is available pursuant to the provisions of Executive Order 11807, and agencies are urged to avail themselves of such assistance.

(**c**) Each agency should seek to utilize the information collected through its management information system to identify unsafe and unhealthful working conditions, and to establish program priorities.

Assistance by the Secretary of Labor in such matters is available to agencies pursuant to the provisions of Executive Order 11807. The guidelines in § 1960.9 of this subpart, which discuss the utilization of agency records and reports, were developed to assist agencies to further joint labor-management safety and health efforts in this regard.

(d) The Department of Labor shall provide Federal agencies with detailed instructions for the proper completion of the recordkeeping and reporting forms specified in §§ 1960.4, 1960.5, 1960.6, and 1960.7, with which agencies shall comply. The Department of Labor shall also provide agencies with sufficient copies of all forms necessary to record and report the required information. Occupational Safety and Health Administration (OSHA) Forms No. 100F, 101F, 102F, 102FF, and instructions for their completion are hereby filed with the Office of the Federal Register as part of the original document. Copies may also be inspected during regular business hours at the Office of Federal Agency Safety Programs, U.S. Department of Labor, Washington, D.C. 20210.

(e) The provisions of this subpart are not intended to discourage agencies from utilizing recordkeeping and reporting forms which contain a more detailed breakdown of information than the forms provided by the Department of Labor, nor are they intended to preclude agencies from establishing accident, injury and illness subcategories within the coded categories established by the Department of Labor for the completion of its forms, provided that subtotals are provided for each coded category established by the Department of Labor.

(f) Information required to be submitted to the Department of Labor by this subpart may be submitted on media processable by electronic data processing equipment provided that such media comply with the requirements of the Office of Federal Agency Safety Programs, U.S. Department of Labor.

(g) Information concerning occupational injuries, illnesses or accidents which, pursuant to statute or Executive Order, must be kept secret in the interest of national defense or foreign policy, shall be recorded on separate forms. Such records shall not be submitted to the U.S. Department of Labor, but may be used by the appropriate

Federal agency in evaluating the agency's program to reduce occupational injuries, illnesses and accidents.

1960.4 Record or Log of Federal Occupational Injuries and Illnesses

(a) Each Federal agency shall maintain a record or log of all recordable occupational injuries and illnesses for each establishment. Where both civilian and noncivilian employees are employed at a single establishment, separate records or logs shall be maintained for each category.

(b) Within 6 working days after receiving information of a recordable occupational injury or illness, appropriate information concerning such injury or illness shall be entered on the record or log. For this purpose, OSHA Form No. 100F, or its equivalent, shall be used and shall be completed in the detail required by that form and the instructions contained therein.

1960.5 Supplementary Record of Federal Occupational Injuries and Illnesses

In addition to the record log of Federal occupational injuries and illnesses provided for under § 1960.4, each Federal agency shall maintain a supplementary record for each occupational injury and illness. The record shall be completed within 6 working days after the receipt of information that a recordable occupational injury or illness has occurred. For this purpose, OSHA Form No. 101F, or Federal Employee's Compensation Forms, or other equivalent forms may be used. OSHA Form No. 101F, or its equivalent, shall be completed in the detail required by the form and the instructions contained therein.

1960.6 Quarterly and Annual Summaries of Federal Occupational Injuries and Illnesses

(a) Each Federal agency, on a calendar year basis, should compile an annual summary of occupational injuries and illness for each establishment, and shall compile both a quarterly and annual summary of occupational injuries and illnesses for each reporting unit. The summaries shall be based on the record or log of Federal occupational injuries and illnesses maintained pursuant to § 1960.4. OSHA Form No. 102F shall be used for these purposes, and shall be completed in the form and detail required by that form and the instructions contained therein.

(**b**) Each agency shall furnish the Department of Labor with a copy of its quarterly and annual summaries compiled on the basis of reporting units. Each quarterly summary and the annual summary of Federal occupational injuries and illnesses shall be completed and forwarded to the Department of Labor no later than 45 calendar days after the close of the applicable reporting period.

1960.7 Quarterly and Annual Summaries of Federal Occupational Accidents

(**a**) Each Federal agency, on a calendar year basis, shall compile both a quarterly and an annual summary of Federal occupational accidents for each reporting unit. OSHA Form No. 102FF shall be used for this purpose, and shall be completed in the form and in the detail required by that form and the instruction contained therein.

(**b**) Each quarterly summary and the annual summary of Federal occupational accidents shall be completed and forwarded to the Department of Labor no later than 45 calendar days after the close of the applicable reporting period.

1960.8 Reporting of Serious Accidents

(**a**) Within 2 working days after the occurrence of an employment accident which is fatal to one or more employees, which results in the hospitalization of five or more employees, or which involves property damage of $100,000 or more, or within 2 working days after the occurrence of a death which is the result of an employment accident, the head of the Federal agency shall report the accident either by telephone or by telegraph to the Secretary of Labor. The report shall relate the circumstances of the accident, any actions taken by the agency regarding the accident, the number of fatalities, and the extent of any injuries. The agency head shall also report any employment accident involving both Federal and non-Federal employees which results in a fatality or the hospitalization of five or more such employees. The Secretary of Labor may require such additional reports, in writing or otherwise, as he deems necessary.

(**b**) Agencies shall construe the term "employment accident" in a liberal manner for the purposes of this section, and shall report such accidents even where there is some doubt as to the relationship between the accident and the "course" or "scope" of employment activities. This requirement is necessary in order that the Secretary of Labor may meet his legal obligations, and the reporting of an accident pursuant to this section therefore does not preclude an agency from making separate determinations regarding the circumstances of the accident as they may relate to administrative or legal proceedings to establish liability for compensation.

1960.9 Location and Utilization of Records and Reports

(**a**) Section 2(b)(13) of the Act declares that one of the purposes of the Act is to encourage joint labor-management efforts to reduce injuries and disease arising out of employment; and, as set forth in § 1960.1(e), the participation of all employees and labor organizations representing employees has been deemed particularly significant in the success of a Federal agency's occupational safety and health program. The provisions of this section, dealing with the availability of information compiled pursuant to the provisions of this subpart, are designed to guide agencies in providing agency employees and their representatives with the basic information necessary to assure that they can actively participate in an agency safety and health program. The provisions of this section are also designed to encourage agencies to allow agency safety and health inspectors to have direct access to the accident, injury and illness records of the establishments they are inspecting in order that they may better carry out their duties pursuant to Subpart D of this part.

(**b**) The log and supplementary records required by §§ 1960.4 and 1960.5 should be maintained at each establishment. Where, for reasons of efficient administration or practicality, an agency must maintain these records at a place other than at each establishment, such agency should ensure that there is available at each establishment a copy of these records. The copy of the log so maintained or made available at an establishment should reflect separately the injury and illness experience of that establishment. These records should be complete and as current as possible; in no case should more than 45 days elapse after the recording of an illness or injury occurring in an establishment and the availability of the records reflecting that injury or illness at that establishment.

(**c**)

1: For agencies engaged in activities such as agriculture, construction, transportation, communications, and electric, gas and sanitary services, which may be physically dispersed, the log and supplementary records, or copies thereof, may be maintained at a place to which employees report each day.

2: For personnel who do not primarily report or work at a single establishment, and who are generally not supervised in their daily work, such as traveling employees, technicians, engineers, etc., the log and supplementary records, or copies thereof, may be maintained at the location from which they are paid or the base from which personnel operate to carry out their activities.

(**d**) Each Federal agency should post a copy of the annual summary of Federal occupational injuries and illnesses for an establishment, as compiled pursuant to § 1960.6, at such establishment, no later than 45 calendar days after the close of the calendar year, or otherwise disseminate a copy of the annual summary for an establishment in written form to all employees of the establishment. Copies of the annual summary should be posted for a minimum of 30 consecutive days in a conspicuous place or places in the establishment where notices to employees are customarily posted. Where establishment activities are physically dispersed, the notice may be posted at the location to which employees report each day. Where employees do not primarily work at or report to a single location, the notice may be posted at the location from which the employees operate to carry out their activities. Each Federal agency should take any necessary steps to ensure that such summary is not altered, defaced, or covered by other material.

(**e**) The head of each agency should make provision to ensure the availability of the records maintained under this subpart to employees and, with the permission of the employees involved, to representatives of employees. Such provision should be in accordance with other applicable statutes and regulations, and any applicable collective bargaining agreements.

(f) Agency safety and health inspectors should have access to accident, injury and illness records in accordance with the provisions of § 1960.26(b).

1960.10 Access to Records by Secretary of Labor

The records required to be maintained under the provisions of this subpart shall be available and made accessible to the Secretary of Labor or his authorized representative (including personnel of the National Institute for Occupational Safety and Health) unless such records are specifically required by statute or Executive Order to be kept secret in the interest of national defense or foreign policy, in which case the Secretary of Labor shall have access to only such information as will not jeopardize national defense or foreign policy. The Secretary of Labor or his authorized representative shall request access to such records from the head of the agency prior to examination.

1960.11 Retention of Records

The records and reports required to be maintained under the provisions of this subpart shall be retained by each agency for 5 years following the end of the calendar year to which they relate, at any location including a Federal record retention center, to which the Secretary of Labor or his authorized representative would have reasonable access.

1960.12 Plan of Action

If it has not already done so by the effective date of this part, each Federal agency shall submit the following information to the Department of Labor no later than January 1, 1975 and at such other times as changes occur:

(a) A list of the names and addresses of each Federal reporting unit which will be covered in the records and reports required by this subpart.

(**b**) The average number of full-time and part-time personnel employed in each reporting unit for which separate records and reports will be maintained.

(**c**) A brief description of any differences between an agency's internal recordkeeping and reporting system and the recordkeeping and reporting system provided by this subpart.

Any Federal agency created or reorganized after October 1, 1974 shall submit an appropriate plan within sixty working days of commencement of operations as a new entity.

1960.13 [Reserved]

1960.14 [Reserved]

SUBPART C—AGENCY ORGANIZATION

1960.15 Purpose and Scope

(**a**) The provisions of this subpart have been developed by the Secretary of Labor to provide guidance to agency heads in the management of an effective and comprehensive occupational safety and health program.

(**b**) Nothing in this subpart is intended in any way to modify the organization or the operation of an agency health services program conducted pursuant to 5 U.S.C. 7901 and OMB Circular A-72. Such a program can, however, contribute to the successful implementation of the occupational safety and health program set forth by this part, by providing for medical examinations, other health monitoring procedures, and the maintenance of medical records, where agency safety and health standards adopted pursuant to this part so require. The Department of Labor will cooperate with the Civil Service

Commission and other appropriate authorities and organizations to resolve problems which arise in connection with the implementation of an agency occupational safety and health program in relation to health services and other personnel policies and programs.

1960.16 Designated Safety and Health Official

(**a**) Executive Order 11807 provides that the head of each agency shall:

> Designate or appoint, to be responsible for the management and administration of the agency occupational safety and health program, an agency official with sufficient authority to represent effectively the interest and support of the agency head.

It is the considered judgment of the Secretary of Labor that an official of the rank of Assistant Secretary, or of equivalent rank or equivalent degree of responsibility, would be of such stature as to be able to fill such a position adequately. It is also the considered judgment of the Secretary of Labor that in order for such official "to represent effectively the interest and support of the agency head," such official should have sufficient headquarters staff with necessary training and experience, and who report directly and exclusively to such official, to carry out his functions under this part.

(**b**) The designated safety and health official should assist the agency head in establishing:

1: an occupational safety and health policy to carry out the provisions of section 19 of the Act of Executive Order 11807;

2: an organization and set of procedures that will effectively implement that policy by observing the provisions of this part, considering the mission, size and organization of the agency;

3: goals and objectives for reducing and eliminating occupational accidents, injuries and illnesses;

4: plans and procedures for evaluating the agency's occupational safety and health program effectiveness at all operational levels; and

5: priorities with respect to the factors which cause occupational accidents, injuries and illnesses so that appropriate corrective action can be taken.

(c) The designated safety and health official should assist the agency head in taking appropriate steps to provide sufficient funds for necessary safety and health staff, equipment, material, and training required to ensure an effective agency occupational safety and health program.

1960.17 Safety and Health Committees

The head of each agency should provide for the establishment of agency safety and health committees, composed of representatives of management and representatives of the employees, at the national level, at the regional or comparable level, and at the establishment level, for the purpose of advising and assisting agency officials, at those respective levels, with respect to their responsibilities under the agency occupational safety and health program. Such committees may also include technical personnel in accordance with the functions to be performed by a particular committee. Suggested functions of such committees are set forth in §§ 1960.25(b) and 1960.41, but these are not exclusive.

1960.18 Posting of Notice; Availability of Act, This Part, and Details of the Agency Safety and Health Program

(a) Each agency should post and keep posted a notice or notices informing employees of the protections and obligations provided for in the Act, Executive Order 11807 and agency programs under this part. The Department of Labor will furnish a uniform poster to agencies. Each agency should add to this uniform poster, or include in its notice or notices, the details of the agency's procedures (established pursuant to § 1960.31) for reports by employees of possible unsafe or unhealthful working conditions of which they have cognizance, the location where employees will be able to obtain information about the agency's occupational safety and health program, including specific agency occupational safety and health standards, and relevant information about any establishment safety and health committee. Such notice or notices should be posted by the agency in each establishment in a conspicuous place or places where notices to employees are customarily posted. Such notices should not be altered, defaced, or covered by other material, and should be kept up to date. Agencies may also convey the information required by this paragraph to employees by other means, provided the notice or notices are also posted in accordance with this paragraph.

(b) Copies of the Act, Executive Order, regulations and guidelines published in this part, details of the agency's safety and health program and applicable safety and health standards, or summaries of any of the forgoing items, should be made available upon request to employees or employee representatives for review in the establishment where the employees are employed as soon as practicable and at a time mutually convenient to the employees and employee representatives and the agency.

1960.19 Duties of Agency Officials and Employees

(a) Each employee who exercises any supervisory functions should comply with agency occupational safety and health standards and all rules, regulations, and orders issued by the head of the agency with respect to the agency occupational safety and health program. In addition, any such employee who is the official in charge of an establishment should comply with any additional rules, regulations or orders issued by the head of the agency to implement the provisions of Subpart D of this part with respect to the particular duties of such an official in the identification and correction of unsafe or unhealthful working conditions.

(b) Each employee should comply with agency occupational safety and health standards and all rules, regulations, and orders issued by the head of the agency which are applicable to an employee's own actions and conduct; and each employee should report any unsafe or unhealthful working condition of which he becomes aware to employees who exercise supervisory functions, or, pursuant to the procedure established in accordance with the provisions of § 1960.31, to agency safety and health officials.

(c) The head of each agency should ensure that in any evaluation of performance or potential, the excellence or culpable failure of each official in charge of an establishment, supervisory employee, or other employee in the performance of his or her occupational safety and health responsibilities be taken into consideration in accordance with any applicable rules of the Civil Service Commission or other appropriate authority. Recognition of group or individual superior performance should be encouraged.

(**d**) The head of each agency should ensure that needed safeguards are included in the agency occupational safety and health program to ensure that no employee is subject to restraint, interference, coercion, discrimination or reprisal by virtue of such employee's participation in the agency occupational safety and health program, including the filing of a report of an unsafe or unhealthful working condition, the initiation of any proceeding under or related to this program, participation by comment or testimony in such proceeding, or the exercise by such employee on behalf of himself or of others of any other right afforded by section 19 of the Act, Executive Order 11807, and the agency program established pursuant to this part. These safeguards should include procedures for the enforcement of these rights which should be consistent with any rules and regulations of the Civil Service Commission and of the agency involved which deal with such matters of restraint, interference, coercion, discrimination or reprisal.

1960.20—24 [Reserved]

SUBPART D—PROCEDURES FOR INSPECTIONS AND ABATEMENTS

1960.25 Purpose, Scope and General Provisions

(**a**) Executive Order 11807 provides that the head of each agency shall:

> * * * assure prompt attention to reports by employees or others of unsafe or unhealthful working conditions; assure

periodic inspections of agency workplaces by personnel with sufficient technical competence to recognize unsafe and unhealthful working conditions in such workplaces; and assure prompt abatement of unsafe or unhealthful working conditions, including those involving facilities and/or equipment furnished by another Government agency, informing the Secretary of significant difficulties encountered in this regard.

The purpose of this subpart is to provide guidance to agency heads in carrying out these duties.

(**b**) It is the general intent of these guidelines that day to day responsibility for the inspection and abatement activities to be carried out pursuant to the provisions of this subpart be delegated by designated safety and health officials of agencies to appropriate agency personnel qualified for this purpose. The Secretary of Labor recognizes, however, that designated safety and health officials may desire and should in fact retain personal responsibility for some day to day agency safety and health activities. Appropriate provisions has therefore been made in these guidelines for the direct exercise of responsibility by designated safety and health officials where, as in §§ 1960.31, 1960.32, and 1960.34, communication with the Secretary of Labor may be involved.

(**c**) Safety and health committees at the establishment and higher levels, as described in § 1960.17, can play a significant role in assisting the designated safety and health official and his respective designees in carrying out their safety and health duties and the provisions of this subpart, as suggested by §§ 1960.26(b), 1960.29(a) and 1960.31(f). Such committees should be kept informed of safety and health matters within their area of concern, as provided in §§ 1960.32, 1960.33(a) and 1960.34(c).

(**d**) The provisions of this subpart are not intended to relieve agencies which occupy space for which the General Services Administration or another agency has assignment responsibility from the duties imposed upon them by such occupancy, including the development and maintenance of sound fire prevention programs for such facilities, the conservation of services and supplies, the use of good housekeeping methods, the preservation of a good working atmosphere, participation in a Facility Self-Protection Plan for dealing with safety emergencies, and payment of user charges. Agencies providing safety and health services pursuant to this subpart and which occupy space for which GSA or another agency has assignment responsibility should take note of those services which GSA or the other agency provides for various levels of user charges, and appropriate reimbursement provisions where the agency performs the services for which GSA or the other agency has responsibility.

(e) Nothing in the provisions of this subpart is intended to preclude arrangements between agencies for the exchange of information and personnel necessary to carry out these provisions.

(f) Executive Order 11807 authorizes assistance to agencies by the Secretary of Labor, upon request and reimbursement for the expenses thereof, in the training of appropriate agency safety and health personnel, the conduct of inspections, and the abatement of unsafe or unhealthful working conditions. Agencies are encouraged to take advantage of such assistance, particularly with respect to the investigation of serious accidents.

(g) The Secretary of Labor has determined that in order to successfully perform his consultation, evaluation and guidance functions pursuant to Section 3 of Executive Order 11807, he needs certain information from each agency about special problems that occur in agency inspection and abatement activities. Accordingly, agencies should furnish the information requested in §§ 1960.31, 1960.32 and 1960.34, pursuant to agreements with the Secretary concerning the transmittal of information.

1960.26 Safety and Health Inspectors; Frequency of Inspection

(a) Executive Order 11807 requires that each agency utilize as inspectors "personnel with sufficient technical competence to recognize unsafe or unhealthful working conditions" in the workplaces to be inspected. For workplaces there is an increased risk of accident, injury or illness due to the nature of the work performed, as in the case of chemical or machine processes or material-handling or loading operations, inspections should therefore be made by a safety and health specialist, as defined in § 1960.2(h) of this part. For workplaces where there is little risk involved, inspections need not be made by a safety and health specialist, but should be conducted by a person having sufficient training and/or experience in the safety health needs of the workplaces involved to adequately carry out the duties of an inspector as set forth in Executive Order 11807 and this subpart. Also inspectors should be accompanied on such inspections by representatives of the official in charge of the establishment being inspected, and representatives of the employees of such establishment, pursuant to the provisions of § 1960.29.

(**b**) Agencies should authorize safety and health inspectors to utilize the services of additional technical and professional personnel, including labor organization and/or safety committee personnel who possess such expertise, to aid them to evaluate the safety and health of working conditions while conducting an inspection. All safety and health inspectors should be provided with technical test equipment where appropriate.

(**c**) Each agency which has areas containing information classified in the interest of national security should provide access to safety and health inspectors who have obtained the appropriate security clearance.

(**d**) All workplaces, including offices, should be inspected at least once annually. For all workplaces where there is an increased risk of accident, injury or illness due to the nature of the work performed, inspections should be conducted more frequently, as determined by the designated safety and health official or his designee based upon extent and degree of risk of accident, injury or illness involved.

1960.27 CONDUCT OF INSPECTION

(**a**) For the purpose of assuring safe and healthful working conditions for employees of agencies, safety and health inspectors should be authorized to enter without delay, and at reasonable times, any building, installation, facility, construction site, or other area, workplace or environment where work is performed by employees of the agency; to inspect and investigate during regular working hours and at other reasonable times, and within reasonable limits and in a reasonable manner, any such place of employment, and all pertinent conditions, structures, machines, apparatus, devices, equipment and materials therein; and to question privately any employee, and/or any supervisory employee, and/or any official in charge of an establishment. Subject to these provisions, and to the provisions of §§ 1960.26 and 1960.28, inspections should take place at such times and in such establishments as the designated safety and health official or his designee directs.

(**b**) Prior to commencement of an inspection, the inspector should be instructed to examine appropriate accident, injury and illness records of the establishment to be inspected, pursuant to § 1960.9(f), in order to facilitate the identification of unsafe or unhealthful working conditions.

(**c**) Safety and health inspectors should be instructed to take environmental samples where appropriate, to take or obtain photographs related to the purpose of the inspection, and to employ other reasonable techniques of inspection.

(**d**) Safety and health inspectors should be instructed to comply with all safety and health rules and practices at the establishment being inspected, and to wear and use appropriate protective clothing and equipment.

(**e**) The conduct of inspections should be such as to preclude unreasonable disruption of the operations of the establishment.

(**f**) At the conclusion of an inspection, the safety and health inspector should confer with the official in charge of the establishment or his representative, and an appropriate representative of the employees of the establishment, and informally advise them of any apparent unsafe or unhealthful working conditions disclosed by the inspection. During such conference, the official in charge of the establishment and the employee representative should be afforded an opportunity to bring to the attention of the safety and health inspector pertinent information regarding conditions in the workplace.

1960.28 Advance Notice of Inspections

(**a**) Advance notice of inspections should not be given to the official in charge of an establishment, except in the following situations:

1: in cases of apparent imminent danger, to enable the official in charge of an establishment to abate the danger as quickly as possible;

2: in circumstances where the inspection can most effectively be conducted after regular business hours or where special preparations are necessary for an inspection;

3: where necessary to assure the presence of representatives of the official charge of the establishment or representatives of employees, or the appropriate personnel needed to aid in the inspection; and

4: where required by security regulations.

(**b**) In the situations described in paragraph (a) of this section, advance notice of inspections should be given only if authorized by the designated safety and health official or his designee, except that in cases of apparent imminent danger, advance notice could be given by the safety and health inspector without such authorization if the designated safety and health official or his designee is not immediately available. When advance notice is given to the official in charge of the establishment, it should be his responsibility to notify promptly, upon receipt of this information, the representative of employees for the purposes set forth in § 1960.29. Advance notice in any of the situations described in paragraph (a) of this section should not be given more than 24 hours before the inspection is scheduled to be conducted, except in unusual circumstances.

1960.29 REPRESENTATIVES OF OFFICIALS IN CHARGE AND REPRESENTATIVES OF EMPLOYEES

(**a**) Safety and health inspectors should be in charge of inspections and questioning of persons. A representative of the official in charge of an establishment and a representative of employees under his supervision should be given an opportunity to accompany the safety and health inspector during the physical inspection of any workplace, both to aid the inspection and to provide such representatives with more detailed knowledge about any existent or potential unsafe or unhealthful working conditions. A safety and health inspector should also arrange for additional representatives of the official in charge and additional representatives of employees to accompany him where he determines that such additional representatives will further aid the inspection. A different representative of the official in charge and a different representative of employees may be allowed to accompany the safety and health inspector during each different phase of an inspection. The members of an establishment's safety and health committee, created pursuant to § 1960.17, may act in the capacity of representatives for the purposes of this section if the committee and the official in charge of the establishment so agree.

(**b**) Safety and health inspectors should be authorized to deny the right of accompaniment under this section to any person whose participation interferes with a fair and orderly inspection. With regard to facilities classified in the interest of national security, only persons authorized to have access to such facilities should be allowed to accompany a safety and health inspector in such areas.

1960.30 Consultation with Employees

Safety and health inspectors should consult with employees concerning matters of occupational safety and health to the extent the inspectors deem necessary for the conduct of an effective and thorough inspection. During the course of an inspection, any employee should be afforded an opportunity to bring to the attention of the safety and health inspector any unsafe or unhealthful working condition which he has reason to believe exists in the workplace.

1960.31 Reports by Employees of Unsafe or Unhealthful Working Conditions

(**a**) The purpose of this section is to provide guidance in the establishment of a channel of communication between agency employees and those with responsibilities for safety and health matters which will assure prompt analysis and response to reports of alleged unsafe or unhealthful working conditions in accordance with the requirements of Executive Order 11807. Since many safety and health problems can be eliminated as soon as they are identified, the existence of this channel of communication is intended to supplement oral reports of unsafe or unhealthful working conditions made by employees to their supervisors, not to act as a substitute for such reports. At the same time, however, an employee should not be required to await the outcome of such an oral report before filing a written report pursuant to the provisions of this section. Nothing in this section is intended to interfere in any way with the prior, simultaneous or subsequent use of any employee of the grievance procedures established pursuant to Executive Order No. 11491, as amended, Executive Order 11636, collective bargaining agreement, or 5 CFR Part 771 (or military equivalent) as a means of requesting correction of alleged unsafe or unhealthful working conditions.

(b) Any employee or representative of employees who believes that an unsafe or unhealthful working condition exists in any workplace where such employee is employed, should be authorized to request an inspection of such workplace by giving notice of the alleged unsafe or unhealthful working condition to the designated safety and health official, or to his designee for this purpose. Any such report should be reduced to writing; should set forth with reasonable particularity the grounds for the report; and should be signed by the employee or representative of employees. Upon the request of the person making such report, the designated safety and health official or his designee for this purpose should not disclose the name of such person or the names of individual employees referred to in the report to anyone other than authorized representatives of the Secretary of Labor, except as provided in paragraph (c) of this section. In the case of imminent danger situations, employees should be allowed to make reports first by telephone or telegraph and reduce them to writing as soon as practicable thereafter.

(c) The designated safety and health official or his designee should consider the report and determine within 5 working days after receipt of such report whether there are reasonable grounds to believe that the alleged unsafe or unhealthful working condition exists. If he does so determine, he should cause an inspection to be made as soon thereafter as possible to determine if such alleged unsafe or unhealthful working condition does in fact exist. If the inspector is unable to locate the alleged unsafe or unhealthful working condition without the assistance of the person who submitted the report, the designated safety and health official or his designee may give the inspector the name of such person, but he should satisfy himself that the name of the person submitting the report and the names of individual employees referred to in the report will not be disclosed to anyone else. In the event the employee report, whether oral or in writing, describes an unsafe or unhealthful working condition which may present imminent danger to the safety or health of employees, the designated safety and health official or his designee should make an immediate determination as to whether there are reasonable grounds to believe that the alleged unsafe or unhealthful working condition exists; and if he does so determine, he should cause an immediate inspection to be made.

(**d**) Inspections initiated pursuant to this section need not be limited to matters referred to in the report of an alleged unsafe or unhealthful working condition. Prior to or during any inspection of a workplace initiated pursuant to this section, any employee employed in such workplace, or representative of employees, should be permitted to notify the safety and health inspector of any other unsafe or unhealthful working condition which he has reason to believe exists in such workplace.

(**e**) If the designated safety and health official or his designee determines that there are no reasonable grounds to believe an unsafe or unhealthful working condition exists, or if an inspection is made on the basis of a report alleging such condition but no such condition is determined to exist, the employee or representative of employees who filed the report should be so notified in writing. The employee or representative of employees should be given an opportunity for prompt and informal review of such determination by appropriate officials, including final review by the designated safety and health official. Any determination made during this review process should be in the form of a written statement setting forth the reasons for such disposition. Employees and employee representatives should be informed of these rights and procedures for review.

(**f**) The designated safety and health official may utilize as his designee for the purposes of this section, where this section entitles a designee to act on his behalf, an appropriate safety and health committee created pursuant to § 1960.17, but he should satisfy himself that the confidentiality of the identity of the persons making or named in a report of an alleged unsafe or unhealthful working condition will be adequately preserved.

(g) Agencies should include in their procedures a means for any employee or representative of employees who filed a report alleging an unsafe or unhealthful working condition, and who is dissatisfied with the final disposition by the agency, to contact in writing the Office of Federal Agency Safety Programs, U.S. Department of Labor (with a copy to the designated safety and health official), describing in detail the entire processing of the report of the alleged unsafe or unhealthful working conditions and setting forth his or her objections thereto. Each such person should be notified of such right by the agency upon final disposition of his report. The Office of Federal Agency Safety Programs, pursuant to § 1960.25(g), may request the agency to submit a report of its investigation, and may arrange for an inspection of the alleged unsafe or unhealthful working condition if necessary. Each agency should maintain its files on such reports and their disposition intact for 5 years following the end of the calendar year to which they relate, at any location, including a Federal record retention center, to which the Secretary of Labor or his authorized representative would have reasonable access.

1960.32 Imminent Danger

Whenever and as soon as a designated safety and health official or his designee concludes on the basis of an inspection that conditions or practices exist in any place of employment which could reasonably be expected to cause death or serious physical harm immediately, or before the imminence of such danger can be eliminated through the normal abatement procedures described in §§ 1960.33 and 1960.34, he should inform the affected employees and official in charge of the establishment of the danger. The official in charge of the establishment, or a person empowered to act for him in his absence, should undertake immediate abatement and the withdrawal of employees not necessary for abatement of the dangerous conditions. In the event the official in charge of the establishment needs assistance to undertake full abatement, he should promptly contact the designated safety and health official and other responsible agency officials, who should assist him. Pursuant to § 1960.25(g), the designated safety and health official should inform the Secretary of Labor, as soon as time permits, of any imminent danger which cannot be promptly and completely abated. Agency safety and health committees should be informed of all relevant actions as soon as time permits, as should representatives of the employees.

1960.33 Notices of Unsafe or Unhealthful Working Conditions

(a) Each agency should establish a procedure for issuing notices of unsafe or unhealthful working conditions discovered upon inspection. Notices should describe with particularity the nature of the unsafe or unhealthful working condition, including a reference to the standard or other requirement involved. The notice should also fix a reasonable time for the abatement of the unsafe or unhealthful working condition. A copy of the notice should be sent to the official in charge of the establishment, and to the safety and health committee of the establishment, if any.

(b) If a notice of an unsafe or unhealthful working condition is issued as a result of a report filed pursuant to § 1960.31, a copy of the notice of the unsafe or unhealthful working condition should also be sent to the person who made such report or notification.

(c) Upon receipt of any notice of an unsafe or unhealthful working condition, the official in charge of an establishment should immediately post such notice, or copy thereof, unedited, at or near each place an unsafe or unhealthful working condition referred to in the notice exists or existed. Where, because of the nature of the establishment operations, it is not practicable to post the notice at or near each such place, such notice should be posted, unedited, in a prominent place where it will be readily observable by all affected employees. For example, where establishment activities are physically dispersed, the notice may be posted at the location to which employees report each day. Where employees do not primarily work at or report to a single location, the notice may be posted at the location from which the employees operate to carry out their activities. The official in charge of an establishment should take steps to ensure that the notice is not altered, defaced, or covered by other material.

(d) Each notice of an unsafe or unhealthful working condition, or a copy thereof, should remain posted until the unsafe or unhealthful working condition has been abated, or for 3 working days, whichever is later.

1960.34 Correction of Unsafe or Unhealthful Working Conditions

(a) The official in charge of an establishment should have primary responsibility for the correction of unsafe or unhealthful working conditions brought to his attention by any means. Where a notice of an unsafe or unhealthful working condition has been issued pursuant to § 1960.33, abatement should be within the time set forth in the notice.

(b) The procedures for correcting unsafe or unhealthful working conditions should include reinspection, where practicable, to determine whether the correction was made. If upon reinspection, it appears that the correction was not made, or was not carried out in accordance with an abatement plan submitted pursuant to paragraph (c) of this section, the designated safety and health official should inform the head of the agency for appropriate action, including action in accordance with the provisions of § 1960.19(c).

(c) The official in charge of the establishment should immediately submit an abatement plan to the designated safety and health official, if in his judgment the abatement of an unsafe or unhealthful working condition will not be possible within 30 working days. Such plan should contain an explanation of the circumstances of the delay in abatement, a proposed timetable for the abatement, and a summary of steps being taken in the interim to protect employees from being injured by the unsafe or unhealthful working condition. A copy of the plan should be sent to the safety and health committee of the establishment, if any, for appropriate comment and assistance. If the estimated abatement time is more than 60 working days, the designated safety and health official shall forward a copy of the plan to the agency head who should convey it to the Secretary of Labor. The head of each agency should inform the Secretary of Labor, pursuant to § 1960.25(g) and at regular intervals to be determined by the Secretary in accordance with the scope and extent of the risk to employee safety and health involved, as to the progress made in carrying out the abatement plan. Any changes in an abatement plan will require the submission of a new plan in accordance with the provisions of this section.

1960.35—39 [Reserved]

SUBPART E—AGENCY OCCUPATIONAL SAFETY AND HEALTH STANDARDS

1960.40 Purpose and Scope

Executive Order 11807 requires that the head of each Federal agency establish procedures for the adoption of agency occupational safety and health standards, and that these agency standards be "consistent" with the standards promulgated by the Secretary of Labor pursuant to section 6 of the Act and applicable to private employment (hereinafter referred to as the Occupational Safety and Health Administration (OSHA) standards). For the purposes of this subpart, standards are "consistent" with OSHA standards if they provide protection to employees which is at least as effective, as the protection provided by the OSHA standards. Executive Order 11807 requires the Secretary of Labor to provide "such consultation to agencies as he deems necessary and appropriate to ensure" that agency standards are consistent with OSHA standards. Specific assistance to agencies in the evaluation of working conditions and the adoption of standards is available from the Secretary of Labor pursuant to the provisions of Executive Order 11807. The purpose of this subpart is to provide guidance to agencies as to all aspects of standards adoption and application, based upon the experience of the Secretary of Labor in this regard, including guidance as to the type of consultation that should be undertaken with the Secretary of Labor prior to the adoption of various types of agency standards. In carrying out his responsibilities under this subpart, the Secretary of Labor shall have due regard for the need of agencies to move promptly in adopting agency safety and health standards.

1960.41 Procedures for Adoption

Executive Order 11807 requires the head of each agency to establish procedures for the adoption of any agency occupational safety and health standards. These procedures should include special provisions for the adoption of emergency temporary agency standards pursuant to § 1960.45, and which parallel those of section 6(c) of the Act. These procedures should also provide an opportunity for written comment by all interested persons. Agency safety and health committees may play an active role in the formulation and consideration of agency occupational safety and health standards, as appropriate. Where employees of one or more Federal agencies primarily report to work in an establishment, as defined in § 1960.2(e), which is physically located on an establishment of another agency, such employees should be considered as interested parties with respect to the actions of such other agency pursuant to this subpart.

1960.42 Initial Adoption of Agency Standards

(**a**) In the order to meet the requirements of Executive Order 11807, agencies should proceed to adopt agency standards as soon as possible, pursuant to the provisions of § 1960.41 and this section.

(**b**) The OSHA standards should in most cases be adopted as agency occupational safety and health standards unless an agency head determines that employees of the agency are not and will not be exposed to working conditions for which an appropriate group of OSHA standards have been promulgated (i.e., that specific subparts of Parts 1910, 1915, 1916, 1917, 1918 and 1926 of this chapter are not relevant to agency working conditions). Consultation with the Secretary of Labor will be available in this regard and will consist of a review of an agency's own evaluation of the nature of agency working conditions, or such additional assistance as is requested by an agency pursuant to the provisions of Executive Order 11807.

(**c**) Where an agency has already adopted, prior to October 1, 1974 comprehensive agency occupational safety and health standards for the protection of agency employees which are not OSHA standards, the head of such agency may request the Secretary of Labor to consult with him as to the appropriateness of readoption of such standards as the agency occupational safety and health standards required to be adopted pursuant to the provisions of this subpart. Such a request should include copies of the standards proposed to be so readopted, arranged insofar as practicable to correspond to appropriate subparts of the OSHA standards contained in Parts 1910, 1915, 1916, 1917, 1918 and 1926 of this Chapter, and should also include any other pertinent information.

(**d**) Agencies which traditionally adopt occupational safety and health standards as, and only as, part of particular job operation descriptions such as technical manuals, rather than as standards of general applicability to all employees, may request the Secretary of Labor to consult with them as to the consistency of such standards with OSHA standards, and as to the appropriateness of adoption of such standards pursuant to the provisions of this part. Such a request should be accompanied by a description of the system utilized and its scope, proposals to assure that such particular standards are and will be as effective as OSHA standards, and proposals to assure the participation of all interested persons in the adoption of such standards. Where the Secretary of Labor is unable to determine whether such standards are fully consistent with OSHA standards, he shall consult with the agency head as to the appropriate steps he believes necessary.

1960.43 Adoption of Different and/or Supplementary Agency Standards

(**a**) The head of an agency, at any time after the adoption of standards pursuant to § 1960.42, may adopt in place of particular standards adopted pursuant to § 1960.42 such different standards as he determines are necessary and appropriate for specialized application to particular working conditions and other related needs of the agency. Such standards shall be consistent with the equivalent OSHA standards in accordance with the provisions of the Executive Order; and the head of such agency should consult with the Secretary of Labor prior to the adoption of such standards so as to allow the Secretary to provide such technical advice and guidance as may be necessary and appropriate in making such a determination.

(b) The head of an agency, at any time, should adopt such supplementary standards as he determines are necessary and appropriate for application to working conditions of agency employees for which there exists no appropriate OSHA standards. The head of such agency should consult with the Secretary of Labor prior to the adoption of such standards so as to allow the Secretary to inform the agency head of any relevant matters of which he is aware.

(c) The head of each agency may revise, modify, or revoke any agency occupational safety or health standard, but such actions should be taken in accordance with the procedures established under § 1960.41 and pursuant to other appropriate provisions of this subpart.

1960.44 Conflicting Standards

(a) Where employees of different agencies primarily engage in joint operations, and/or primarily report to work or carry out operations in the same establishment, as defined in § 1960.2(e), the heads of the agencies involved should consult with each other and with the Secretary of Labor as to the resolution of any conflict or potential conflict between the occupational safety and health standards of the agencies for the conduct of such joint operations and/or the design of such facilities.

(b) Where the head of an agency is required by law to comply with requirements promulgated by a Federal authority affecting the occupational safety and health of the employees of his agency, such requirements might conflict with the agency occupational safety and health standards adopted pursuant to this subpart; that is, compliance with such requirement may make simultaneous compliance with an agency occupational safety and health standard impossible. For example, standards issued by the General Services Administration pertaining to space for which it has assignment responsibility, pursuant to its statutory authority to conserve and protect such property, might create a conflict with the standards adopted pursuant to this part because GSA standards pertain to certain aspects of fire safety and sanitation, as well as levels of illumination, heating, cooling, and gas consumption for government vehicles. In cases where compliance with standards of another agency conflicts with the duty imposed upon the head of an agency to assure employee safety and health pursuant to Section 19 of the Act and Executive Order 11807, the head of such agency should inform the head of the other Federal authority and the Secretary of Labor of such conflict, so that joint efforts to resolve the conflict may be undertaken.

(c) Appropriate employee representatives should be kept informed of any activities undertaken pursuant to this section.

1960.45 Emergency Standards

(a) In emergency situations, the Secretary of Labor will not have the time necessary to consider whether or not emergency temporary occupational safety and health standards adopted by agencies are "consistent" with those emergency temporary OSHA standards promulgated by the Secretary of Labor pursuant to section 6(c) of the Act. Therefore, in the event the Secretary of Labor does promulgate such a standard, the head of each agency should adopt it without change, and should immediately assure that any agency employees exposed to the unsafe or unhealthful working condition involved receives the protection provided for in such standard unless an emergency affecting the national defense makes this impossible. Such standard should remain effective as an agency standard until such time as the Secretary of Labor promulgates a permanent standard and the agency has completed procedures provided for by this subpart for the adoption of an agency occupational safety and health standard.

(b) An agency head may also adopt emergency temporary agency occupational safety and health standards when he deems such action necessary for the protection of agency employees from grave dangers. Such agency head should immediately inform the authorized representatives of employees of the agency and the Secretary of Labor of such action.

1960.46 Access to Standards

(a) Each agency should notify the Secretary of Labor on a quarterly basis of the final adoption, revision, modification, or revocation of any agency occupational safety and health standard taken within the current quarter, and make copies available to him upon request.

(b) Where any incorporation by references is involved in promulgating, revising or modifying any standard pursuant to this subpart, agencies should follow the rules set forth in I CFR 51.6, 51.7, and 51.8. Difficulties in this regard should be reported to the Secretary of Labor, who will consult with the Director of the FEDERAL REGISTER and then advise agencies in this regard.

1960.47—49 [Reserved]

SUBPART F—FIELD FEDERAL SAFETY AND HEALTH COUNCILS

1960.50 PURPOSE AND SCOPE

Executive Order 11807 provides that the Secretary of Labor shall "facilitate the exchange of ideas and information throughout the Government with respect to matters of occupational and health through such arrangements as he deems appropriate." The Secretary of Labor will establish and continue Field Federal Safety and Health Councils in the fulfillment of this provision with respect to matters of occupational safety and health on a local level. The councils will consist of representatives of local area Federal agencies, and of labor organizations representing employees of local area Federal agencies. The Secretary of Labor will provide leadership and guidance to the Field Federal Safety and Health Councils in fulfilling their responsibilities, and agency heads should ensure that field units within an agency are officially represented and actively participate in the programs of these councils.

1960.51—59 [Reserved]

INSTRUCTIONS FOR COMPLETING LOG OF FEDERAL OCCUPATIONAL INJURIES AND ILLNESSES (OSHA FORM NO. 100F)

Column 1—CASE OR FILE NUMBER

Any number may be entered which will facilitate comparison with supplementary records.

Column 2—DATE OF INJURY OR ILLNESS

For occupational injuries enter the date of the work accident which resulted in injury. For occupational illnesses enter the date of initial diagnosis of illness, or, if absence occurred before diagnosis, the first day of the absence in connection with which the case was diagnosed.

Column 3—EMPLOYEE'S NAME

Column 4—OCCUPATION

Enter the occupational title of the job to which the employee was assigned at the time of injury or illness. In the absence of a formal occupational title, enter a brief description of the duties of the employee.

Column 5—DEPARTMENT

Enter the name of the department to which employee was assigned at the time of injury or illness, whether or not employee was actually working in that department at the time. In the absence of formal department titles, enter a brief description of normal workplace to which employee is assigned.

Column 6—NATURE OF INJURY OR ILLNESS AND PART(S) OF BODY AFFECTED

Enter a brief description of the injury or illness and indicate the part or parts of body affected. Where entire body is affected, the entry "body" can be used.

Column 7—INJURY OR ILLNESS CODE

Enter the one code which most accurately describes the nature of injury or illness. A list of codes appears at the bottom of the log. A more complete description of occupational injuries and illnesses appears below in "definitions."

Column 8—FATALITIES

If the occupational injury or illness resulted in death, enter date of death.

Column 9—LOST WORKDAYS

Enter the number of days the employee would have worked but could not because of occupational injury or illness. The number of lost workdays should not include the day of injury. The number of days includes all days (consecutive or not) on which, because of the injury or illness:

1) the employee would have worked but could not, or
2) the employee was assigned to a temporary job, or
3) the employee worked at a permanent job less than full time, or
4) the employee worked at a permanently assigned job but could not perform all duties normally assigned to it.

For employees not having a regularly scheduled shift, i.e., certain truck drivers, construction workers, part-time employees, etc., it may be necessary to estimate the number of lost workdays. Estimates of lost workdays shall be based on prior work history of the employee and days worked by employees, not ill or injured, working in the department and/or occupation of the ill or injured employee.

Column 10—PERMANENT TRANSFER TO ANOTHER JOB OR TERMINATION OF EMPLOYMENT AFTER LOST WORKDAYS

Complete only if the employee did not return to his previous assignment after lost workdays.

Column 11—NONFATAL CASES WITHOUT LOST WORKDAYS

Enter a check in Column 11 for all cases of occupational injury or illness which did not involve fatalities or lost workdays but did result in:

—Transfer to another job or termination of employment, or
—Medical treatment, other than first aid, or
—Diagnosis of occupational illness, or
—Loss of consciousness, or
—Restriction of work or motion.

Column 12—TRANSFER TO ANOTHER JOB OR TERMINATION OF EMPLOYMENT WITHOUT LOST WORKDAYS

If the check in Column 11 represented a transfer to another job or termination of employment with no lost workdays, enter another check in Column 12.

INITIALING REQUIREMENT

Each line entry regarding an occupational injury or illness must be initialed in the right hand margin by the person responsible for the accuracy of the entry. Changes in an entry also must be initialed in the affected column.

CHANGES IN EXTENT OF OR OUTCOME OF INJURY OR ILLNESS

If there is a change in an occupational injury or illness case which affects entries in Column 9, 10, 11, or 12, the first entry should be lined out and a new entry made. For example, if an injured employee at first required only medical treatment but later lost workdays, the check in Column 11 should be lined out and the number of lost workdays entered in Column 9.

In another example, if an employee with an occupational illness lost workdays, returned to work, and then died of the illness, the workdays noted in Column 9 should be lined out and the date of death entered in Column 8.

An entry may be lined out if later found to be a nonoccupational injury or illness.

DEFINITIONS OF TERMS FOR USE IN RECORDING FEDERAL OCCUPATIONAL INJURIES AND ILLNESSES

OCCUPATIONAL INJURY is any injury such as a cut, fracture, sprain, amputation, etc., which results from a work accident or from exposure in the work environment.

OCCUPATIONAL ILLNESS of an employee is any abnormal condition or disorder, other than one resulting from an occupational injury, caused by exposure to environmental factors associated with his employment. It includes acute and chronic illnesses or diseases which may be caused by inhalation, absorption, ingestion, or direct contact, and which can be included in the categories listed below.

The following listing gives the categories of occupational illnesses and disorders that will be utilized for the purpose of classifying recordable illnesses. The identifying codes are those to be used in Column 7 of the log. For purposes of information, examples of each category are given. These are typical examples, however, and are not to be considered to be the complete listing of the types of illnesses and disorders that are to be counted under each category.

(21) Occupational Skin Diseases or Disorders
Examples: Contact dermatitis, eczema, or rash caused by primary irritants and sensitizers or poisonous plants; oil acne; chrome ulcers; chemical burns or inflammations; etc.

(22) Dust Diseases of the Lungs (Pneumoconioses)
Examples: Silicosis, asbestosis, coal worker's pneumoconiosis, byssinosis, and other pneumoconioses.

(23) Respiratory Conditions Due to Toxic Agents
Examples: Pneumonitis, pharyngitis, rhinitis or acute congestion due to chemicals, dusts, gases, or fumes; farmer's lung; etc.

(24) Poisoning (Systemic Effects of Toxic Materials)
Examples: Poisoning by lead, mercury, cadmium, arsenic, or other metals; poisoning by carbon monoxide, hydrogen sulfide or other gases; poisoning by benzol, carbon tetrachloride, or other organic solvents; poisoning by insecticide sprays such as parathion, lead arsenate; poisoning by other chemicals such as formaldehyde, plastics and resins, etc.

(25) Disorders Due to Physical Agents (Other Than Toxic Materials)
Example: Heatstroke, sunstroke, heat exhaustion and other effects of environmental heat; freezing, frostbite and effects of exposure to low temperatures; caisson disease; effects of ionizing radiation (isotopes, X-rays, radium); effects of nonionizing radiation (welding flash, ultraviolet rays, microwaves, sunburn), etc.

(26) Disorders Due to Repeated Trauma
Examples: Noise-induced hearing loss; synovitis, tenosynovitis, and bursitis; Raynaud's phenomena; and other conditions due to repeated motion, vibration or pressure.

(29) All Other Occupational Illnesses
Examples: Anthrax, brucellosis, infectious hepatitis, malignant and benign tumors, food poisoning, histoplasmosis, coccidioidomycosis, etc.

RECORDABLE OCCUPATIONAL INJURIES AND ILLNESSES are any occupational injuries or illnesses which result in:

1) FATALITIES, regardless of the time between the injury and death, or the length of the illness; or
2) LOST WORKDAYS CASES, other than fatalities that result in lost workdays; or
3) NONFATAL CASES WITHOUT LOST WORKDAYS, which result in transfer to another job or termination of employment, or require medical treatment (as defined below), or involve loss of consciousness or restriction of work or motion. This category also includes any diagnosed occupational illnesses which are reported to the Agency but are not classified as fatalities or lost workday cases.

MEDICAL TREATMENT includes treatment administered by a physician or by registered professional personnel under the standing orders of a physician. Medical treatment does NOT include first aid treatment (one-time treatment and subsequent observation of minor scratches, cuts, burns, splinters, and so forth, which do not ordinarily require medical care) even though provided by a physician or registered professional personnel.

ESTABLISHMENT: A single physical location where business is conducted or where services or industrial operations are performed. (For example: warehouse, or central administrative office.) Where distinctly separate activities are performed at a single physical location (such as contract construction activities operated from the same physical location as a lumber yard), each activity shall be treated as a separate establishment.

For agencies engaged in activities such as agriculture, construction, transportation, communications, and electric, gas and sanitary services, which may be physically dispersed, records may be maintained at a place to which employees report each day.

Records for personnel who do not primarily report to work at a single establishment, such as traveling technicians, engineers, etc., shall be maintained at the location from which they are paid or the base from which personnel operate to carry out their activities.

WORK ENVIRONMENT is comprised of the physical location, equipment, materials processed or used, and the kinds of operations performed by an employee in the performance of his work, whether on or off the Agency's premises.

Nov. 1, 1971
OSHA NO. 100F

LOG OF FEDERAL OCCUPATIONAL INJURIES AND ILLNESSES

This is the separate log for: Civilian Personnel ☐; Military (Non-combat) Personnel ☐.

Case or file no.	Date of injury or initial diagnosis of illness. If diagnosis of illness was made after first day of absence enter first day of absence. (mo./day/yr.)	Employee's Name (First name, middle initial, last name)	Occupation of injured employee at time of injury or illness	Department to which employee was assigned at time of injury or illness	DESCRIPTION Nature of injury or ill body af (Typical entries for th Amputation of 1st jo Strain of lower back Contact dermatitis o Electrocution—body
1	2	3		4	5

(Parent Agency and Federal Establishment Code)

(Sub-Agency & Federal Establishment Code)

(Address of Sub-Agency)

Injury Code
10 All occupational injuries

Reduced Size Facsimile

LOG OF FEDERAL OCCUPATIONAL INJURIES AND ILLNESSES

TION OF INJURY OR ILLNESS		EXTENT OF AND OUTCOME OF INJURY OR ILLNESS					
		Fatalities	Lost Workday Cases		Nonfatal Cases Without Lost Workdays		
y or illness and part(s) of ody affected for this column might be: 1st joint right forefinger r back titis on both hands —body)	Injury or illness code *See codes at bottom of page.*	Enter date of death (mo./day/yr.)	Enter workdays lost due to injury or illness (*see instructions on back.*)	If, *after* lost workdays, the employee was permanently transferred to another job or was terminated, enter a check in the column below	If no entry was made in columns 8 or 9, but the injury or illness did result in: Transfer to another job or termination, or; medical treatment, other than first aid, or; diagnosis of occupational illness, or; loss of consciousness, or; restriction of work or motion; Enter a check in the column below		If a check in column 11 represented a transfer or termination, enter another check in column 12
	6	7	8	9	10	11	12

Illness Codes

21 Occupational skin diseases or disorders
22 Dust diseases of the lungs (pneumoconioses)
23 Respiratory conditions due to toxic agents
24 Poisoning (Systemic effects of toxic materials)
25 Disorders due to physical agents (other than toxic materials)
26 Disorders due to repeated trauma
29 All other occupational illnesses

Reduced Size Facsimile

OSHA NO. 101F
CASE OR FILE NO. ——————

SUPPLEMENTARY RECORD OF FEDERAL OCCUPATIONAL INJURIES AND ILLNESSES

AGENCY

1. Name ....
2. Mail address ....
   (No. and street) (City or town) (State)
3. Location, if different from mail address ....

INJURED OR ILL EMPLOYEE

4. Name .... Social Security No. ....
   (First) (Middle) (Last)
5. Home address ....
   (No. and street) (City or town) (State)
6. Age .... 7. Sex: Male........ Female........ (Check one)
8. Occupation ....
   (Enter regular job title, *not* the specific activity he was performing at time of injury.)
9. Department ....
   (Enter name of department or division in which the injured person is regularly employed, even though he may have been temporarily working in another department at the time of injury.)

THE ACCIDENT OR EXPOSURE TO OCCUPATIONAL ILLNESS

10. Place of accident or exposure ....
    (No. and street)

....
(City or town) (State)

If accident or exposure occurred on Agency's premises, give address of plant or establishment in which it occurred. Do not indicate department or division within the plant or establishment. If accident occurred outside Agency's premises at an identifiable address, give that address. If it occurred on a public highway or at any other place which cannot be identified by number and street, please provide place references locating the place of injury as accurately as possible.

11. Was place of accident or exposure on Agency's premises? ......

....
(Yes or No)

12. What was the employee doing when injured? ....
(Be specific. If he was
....
using tools or equipment or handling material, name them
....
and tell what he was doing with them.)

13. How did the accident occur? ....
(Describe fully the events which resulted
....
in the injury or occupational illness. Tell what happened and how it happened.
....
Name any objects or substances involved and tell how they were involved. Give
....
full details on all factors which led or contributed to the accident. Use separate
....
sheet for additional space.)

OCCUPATIONAL INJURY OR OCCUPATIONAL ILLNESS

14. Describe the injury or illness in detail and indicate the part of body affected ....
(e.g.: amputation of right index finger at second joint;
....
fracture of ribs; lead poisoning; dermatitis of left hand, etc.)

15. Name the object or substance which directly injured the employee. (For example, the machine or thing he struck against or which struck him; the vapor or poison he inhaled or swallowed; the chemical or radiation which irritated his skin, or in cases of strains, hernias, etc., the thing he was lifting, pulling, etc.)
....
....

16. Date of injury or initial diagnosis of occupational illness ....
....
(Date)

17. Did employee die? .... (Yes or No)

OTHER

18. Name and address of physician ....
....

19. If hospitalized, name and address of hospital ....
....

Date of report .... Prepared by ....
Official position ....

SUPPLEMENTARY RECORD OF FEDERAL OCCUPATIONAL INJURIES AND ILLNESSES (OSHA FORM NO. 101F)

To supplement the Log of Federal Occupational Injuries and Illnesses (OSHA No. 100F), reporting activities may find it useful to maintain a record of each recordable occupational injury or illness. Bureau of Employees' Compensation, insurance, or other reports are just as useful as records if they contain all facts listed below or are supplemented to do so.

Such records should contain at least the following facts:

1) About the Agency—Unit's name, mail address, and location if different from mail address.

2) About the injured or ill employee—name, social security number, home address, age, sex, occupation, and department.

3) About the accident or exposure to occupational illness—place of accident or exposure, whether it was on Agency's premises, what the employee was doing when injured, and how the accident occurred.

4) About the occupational injury or occupational illness—description of the injury or illness, including part of body affected; name of the object or substance which directly injured the employee; and date of injury or initial diagnosis of illness.

5) Other—name and address of physician; if hospitalized, name and address of hospital; date of report; and name and position of person preparing the report.

SEE DEFINITIONS ON THE BACK OF OSHA FORM 100F

OSHA No. 102F
Nov. 1, 1971

SUMMARY REPORT OF FEDERAL OCCUPATIONAL INJURIES AND ILLNESSES

A. This is the separate summary report for:

A.1 Civilian Personnel – ☐

A.2 Military (Non-combat) Personnel – ☐

B. Reporting Period ☐☐-☐☐-☐☐
Month Day Year
(Quarter/Year Ending Date)

C. ______________ ☐☐☐☐
(Name of Parent Agency–Federal Establishment Code)

D. ______________ ☐☐☐☐
(Name of Sub-Agency–Federal Establishment Code)

E. ______________ ☐☐☐☐☐
(Address: No. & St., City, Co., State, Zip Code)

Injury and Illness Category		Fatalities	Lost Workday Cases			Nonfatal Cases Without Lost Workdays*	
Code 1	Category 2	3	Number of Cases 4	Number of Cases Involving Permanent Transfer to Another Job or Termination of Employment 5	Number of Lost Workdays 6	Number of Cases 7	Number of Cases Involving Transfer to Another Job or Termination of Employment 8
10	Occupational Injuries						
	Occupational Illnesses						
21	Occupational Skin Diseases or Disorders						
22	Dust diseases of the lungs (pneumoconioses)						
23	Respiratory conditions due to toxic agents						
24	Poisoning (systemic effects of toxic materials)						
25	Disorders due to physical agents (other than toxic materials)						
26	Disorders due to repeated trauma						
29	All other occupational illnesses						
30	Total–occupational illnesses (21-29)						
31	Total–occupational injuries and illnesses (10 plus 30)						
40	Total Man-hours worked by all employees ☐☐☐☐☐☐☐☐ (This Reporting Period)						
50	Average number of employees ☐☐☐☐☐☐☐ (This Reporting Period)						

*Nonfatal Cases Without Lost Workdays–Cases resulting in: Medical treatment beyond first aid, diagnosis of occupational illness, loss of consciousness, restriction of work or motion, or transfer to another job (without lost workdays).

Reduced Size Facsimile

INSTRUCTIONS FOR REPORT PREPARATION OSHA FORM NO. 102F

Insert a check-mark (√) in the appropriate square box to identify data contained in the report as either civilian or military. Please do *not* combine civilian and military data—submit separate reports for each group.

Reporting Period. Underline the appropriate reporting period and insert the last calendar day for which the report covers. The month, day, and year should be recorded numerically in the square boxes such as
[0] [3] - [3] [1] - [7] [2]

Name and code of agency and name and code of unit in that agency. Refer to attached extract from Federal Establishment Code Manual and insert the given code number for the reporting agency in the specified square boxes.

Address. Insert the address of the reporting work-place location which should include the Street number, name of the Street, Avenue, etc., City or Town, County, State, and Zip Code Number. The Zip Code Number should be inserted in the specified square boxes.

(Columns 1 and 2). Self-explanatory.

(Column 3) Fatalities. This represents Federal employee deaths resulting from an occupational injury or illness, regardless of the time between the injury and death, or the length of illness.

(Column 4) Lost workday cases. Injuries and illnesses other than fatalities resulting in lost workday.

(Column 5) Lost workdays involving transfer or termination of employment. Insert total number of cases on the respective code line which represents lost work cases of injured or ill employees resulting in transfers to another job or termination of employment.

(Column 6) Lost workdays. Other than fatalities, lost workdays should be computed as the actual number of days the Federal employee(s) would have worked but could not because of an occupational injury or illness, i.e., an occupational injury or illness that prevented the Federal employee from performing his normal assignment. For persons still absent at end of a reporting period, estimate the expected additional number of lost workdays and include those for the quarter being reported. When the annual report is prepared, the *actual* number of days lost will be entered instead of the estimated figure except for those persons still absent at the end of the annual reporting period. For those still absent at the end of the reporting period, estimate expected additional lost days and include in total for the year.

(Column 7) Non-fatal cases without lost workdays. Insert the total number of cases resulting in: Medical treatment beyond first aid, diag-

nosis of occupational illness, loss of consciousness, restriction of work or motion, or transfer to another job (without the lost workdays).

(Column 8) Non-fatal cases without lost workdays involving transfer or termination of employment. Insert total number of cases on the respective code line which represents cases of a non-fatal type without loss of work that resulted in transfer or termination of employment.

(Code 40) Man-hours worked. Insert the total hours worked by all employees on official duty at the reporting workplace during the reporting period (quarter, annual), excluding vacations, holidays, sick leave, and other nonwork time. Count only the actual hours of overtime worked. If any employee worked irregular hours, or if any part-time workers were employed, care should be taken to include their actual hours worked. Do *not* combine civilian and military man-hours worked. Please do *NOT* report man-days; all man-days should be *converted* to man-hours by the reporting agency.

(Code 50) Average number of employees. Insert the average number of full and part-time employees during the reporting period. Include all classes of employees (i.e., administrative, supervisory, clerical, professional, non-professional, technical, operating related workers, etc.). Do *not* combine civilian and miliary average number of employees.

Method for Recording Data. Enter numbers requested for Codes 40 and 50 in the boxes provided—right to left. For example: Total man-hours worked by all employees for the reporting period are 80,912 hours —the correct entry would be ☐ ☐ [8] [0] [9] [1] [2] An entry should *not* be made as follows:

☐ [8] [0] [9] [1] [2] ☐

All entries should be made to the nearest whole number.

OSHA No. 102FF

SUMMARY REPORT OF FEDERAL OCCUPATIONAL ACCIDENTS

A. This is the separate summary report for:

A.1 Civilian Personnel – ☐

A.2 Military (Non-combat) Personnel – ☐

B. Reporting Period ☐☐-☐☐-☐☐
Month Day Year
(Quarter/Year Ending Date)

C. ______________________ ☐☐☐☐
(Name of Parent Agency–Federal Establishment Code)

D. ______________________ ☐☐☐☐
(Name of Sub-Agency–Federal Establishment Code)

E. ______________________ ☐☐☐☐☐
(Address: No. & St., City, Co., State, Zip Code)

	1.0 Automobiles		2.0 Cranes, Lifts, Etc.,	3.0 Marine	4.0 Aircraft	5.0 Accidents Other Than Vehicles	6.0 Fire	7.0 Tort Claims (Dollars)
	1.1 Gov't	1.2 Private						
8.0 Total Accidents								
9.0 *Vehicle Usage:* 9.1–Total Miles Traveled								
9.2–Total Hours Operated								
10.0 Cost of Repair and/or Replacement–Direct Dollars								

Reduced Size Facsimile

INSTRUCTIONS FOR REPORT PREPARATION OSHA FORM NO. 102FF

An *accident* is defined for OSHA No. 102FF reporting purposes as an unintended or unplanned occurrence that results in injury to personnel, property damage, production interference, or a combination of these conditions.

A *vehicle accident* is any occurrence involving a Federal Government-owned, leased, or rented vehicle, or privately-owned vehicle while operated on official Federal Government business which results in death, injury or property damage of one-hundred dollars ($100) or more, regardless of who was injured (if anyone) or what property was damaged.

Insert a check-mark (√) in the appropriate square box to identify data contained in the report as either civilian or military. Please do *not* combine civilian and military data—submit separate reports for each group.

Reporting Period. Underline the appropriate reporting period and insert the last calendar day the report covers. The month, day, and year should be recorded numerically in the square boxes such as
[0] [3] - [3] [1] - [7] [2]

Name and code of agency and name and code of unit in that agency. Refer to attached Extract from Federal Establishment Code Manual and insert the given code number for the reporting agency in the specified square boxes.

Address. Insert the address of the reporting work-place location which should include the Street number, name of the Street, Avenue, etc., City or Town, County, State, and Zip Code Number. The Zip Code Number should be inserted in the specified square boxes.

1.0 *Automobiles*—means cars, trucks, or motorcycles used for official Federal Government business.

2.0 *Crane, Lifts, etc.*—means normal construction, warehouse, supply room, or yard vehicles operated by Federal employees during the official workday. These vehicles are defined in OSHA Occupational Safety and Health Standards.

3.0 *Marine*—means any water-borne craft, motorized, non-motorized, steam, sail, towed, etc. A vessel capable of being used as a means of transportation on water, including special purpose floating structures not primarily designed for, or used as a means of, transportation on water.

4.0 *Aircraft*—means any air-borne craft, whether powered, towed, or free flying.

5.0 *Accidents Other Than Vehicles*—includes such incidents as the dropping of a typewriter which causes injury to an individual, or the bursting of a water pipe which damages supplies.

6.0 *Fire*—incidents resulting from trash fires, wastepaper basket fires, etc.

7.0 *Tort Claims (Dollars)*—means any wrong resulting from an occupational accident other than a breach of contract for which the laws allow recovery. The total dollar amount paid during the reporting period should be reported.

8.0 *Total Accidents*—Insert in the appropriate column the total number of accidents that occurred during the reporting period.

9.0 *Vehicle Usage*—Insert in the appropriate column the total number of miles traveled/hours operated by all vehicles for the respective reporting period, e.g., an agency has 10 cars which traveled 500 miles each during the reporting period (January 1, 1972 through March 21, 1972). 5,000 miles would be reported in Column 1.1 "Government." This is needed to determine the vehicle accident frequency rate.

10.0 *Direct Dollars: Cost of Repair and/or Replacement*—Insert the dollar cost of repair or replacement of any property damaged as the result of an accident which amounts to $100 or more.

14

ADDITIONAL APPLICABLE PROVISIONS

ADDITIONAL APPLICABLE PROVISIONS

The following additional provisions are included to more thoroughly explain the OSHA law to help ensure compliance.

NATIONAL ADVISORY COMMITTEE

A National Advisory Committee on Occupational Safety and Health consisting of twelve members is provided to advise, consult with, and make recommendations on matters relating to the administration of the act.

"AS STATED"
Sec. 7. "THE ACT"
Sec. 7.(a)(1) *There is hereby established a National Advisory Committee on Occupational Safety and Health consisting of twelve members appointed by the Secretary, four of whom are to be designated by the Secretary of Health, Education, and Welfare, without regard to the provisions of Title 5, United States Code, governing appointments in the competitive service, and composed of representatives of management, labor, occupational safety and occupational health professions, and of the public. The Secretary shall designate one of the public members as Chairman. The members shall be selected upon the basis of their experience and competence in the field of occupational safety and health.*

Sec. 7.(a)(2) *The Committee shall advise, consult with, and make recommendations to the Secretary and the Secretary of Health, Education, and Welfare on matters relating to the administration of the Act. The Committee shall hold no fewer than two meetings during each calendar year. All meetings of the Committee shall be open to the public and a transcript shall be kept and made available for public inspection.*

Sec. 7.(a)(3) *The members of the Committee shall be*

compensated in accordance with the provisions of Section 3109 of Title 5, United States Code.

Sec. 7.(a)(4) *The Secretary shall furnish to the Committee an executive secretary and such secretarial, clerical, and other services as are deemed necessary to the conduct of its business.*

STANDARD-SETTING ADVISORY COMMITTEE

A standard-setting advisory committee of not more than fifteen members, made up of qualified members from management, labor, state agencies, safety and health professional organizations, and standard-producing organizations, is provided to assist in the standard-setting functions.

"AS STATED"
Sec. 7. "THE ACT"
Sec. 7.(b) *An advisory committee may be appointed by the Secretary to assist him in his standard-setting functions under Section 6 of this Act. Each such committee shall consist of not more than fifteen members and shall include as a member one or more designees of the Secretary of Health, Education, and Welfare, and shall include among its members an equal number of persons qualified by experience and affiliation to present the viewpoint of the employers involved, and of persons similarly qualified to present the viewpoint of the workers involved, as well as one or more representatives of health and safety agencies of the States. An advisory committee may also include such other persons as the Secretary may appoint who are qualified by knowledge and experience to make a useful contribution to the work of such committee, including one or more representatives of professional organizations of technicians or professionals specializing in occupational safety or health, and one or more representatives of nationally recognized standards-producing organizations, but the number of persons so appointed to any such advisory com-*

mittee shall not exceed the number appointed to such committee as representatives of Federal and State agencies. Persons appointed to advisory committees from private life shall be compensated in the same manner as consultants or experts under Section 3109 of Title 5, United States Code. The Secretary shall pay to any State which is the employer of a member of such a committee who is a representative of the health or safety agency of that State reimbursement sufficient to cover the actual cost to the State resulting from such representative's membership on such committee. Any meeting of such committee shall be open to the public and an accurate record shall be kept and made available to the public. No member of such committee (other than representatives of employers and employees) shall have an economic interest in any proposed rule.

ASSISTANCE FOR THE SECRETARY OF LABOR

The secretary is authorized to use (with consent) any federal or state agency, and to employ experts and consultants.

"AS STATED"
Sec. 7 "THE ACT"
Sec. 7.(c) *In carrying out his responsibilities under this Act, the Secretary is authorized to—*

(1) use, with the consent of any Federal Agency, the services, facilities, and personnel of such agency, with or without reimbursement, and with the consent of any State or political subdivision thereof, accept and use the services, facilities, and personnel of any agency of such State or subdivision with reimbursement; and

(2) employ experts and consultants or organizations thereof as authorized by Section 3109 of Title 5, United States Code, except that contracts for such employment may be renewed annually; compensate individuals so employed at rates not in excess of the rate specified at the time of

service for grade GS-18 under Section 5332 of Title 5, United States Code, including traveltime, and allow them while away from their homes or regular places of business, travel expenses (including per diem in lieu of subsistence) as authorized by Section 5703 of Title 5, United States Code, for persons in the Government service employed intermittently, while so employed.

OCCUPATIONAL SAFETY AND HEALTH REVIEW COMMISSION

The Occupational Safety and Health Review Commission consists of three members appointed by the president and shall function as follows:

"AS STATED"
Sec. 12. "THE ACT"
Sec. 12.(a) *The Occupational Safety and Health Review Commission is hereby established. The Commission shall be composed of three members who shall be appointed by the President, by and with the advice and consent of the Senate, from among persons who by reason of training, education, or experience are qualified to carry out the functions of the Commission under this Act. The President shall designate one of the members of the Commission to serve as Chairman.*

Sec. 12.(b) *The terms of members of the Commission shall be six years except that (1) the members of the Commission first taking office shall serve, as designated by the President at the time of appointment, one for a term of two years, one for a term of four years, and one for a term of six years, and (2) a vacancy caused by the death, resignation, or removal of a member prior to the expiration of the term for which he was appointed shall be filled only for the remainder of such unexpired term. A member of the Commission may be removed by the President for inefficiency, neglect of duty, or malfeasance in office.*

Sec. 12.(c) *(1) Section 5314 of Title 5, United States Code, is amended by adding at the end thereof the following new paragraph:*

"(57) Chairman, Occupational Safety and Health Review Commission."

(2) Section 5315 of Title 5, United States Code, is amended by adding at the end thereof the following new paragraph:

"(94) Members, Occupational Safety and Health Review Commission."

Sec. 12.(d) *The principal office of the Commission shall be in the District of Columbia. Whenever the Commission deems that the convenience of the public or of the parties may be promoted, or delay or expense may be minimized, it may hold hearings or conduct other proceedings at any other place.*

Sec. 12.(e) *The Chairman shall be responsible on behalf of the Commission for the administrative operations of the Commission and shall appoint such hearing examiners and other employees as he deems necessary to assist in the performance of the Commission's functions and to fix their compensation in accordance with the provisions of Chapter 51 and Subchapter III of Chapter 53 of Title 5, United States Code, relating to classification and General Schedule pay rates: Provided, That assignment, removal, and compensation of hearing examiners shall be in accordance with Sections 3105, 3344, 5362, and 7521 of Title 5, United States Code.*

Sec. 12.(f) *For the purpose of carrying out its functions under this Act, two members of the Commission shall constitute a quorum and official action can be taken only on the affirmative vote of at least two members.*

Sec. 12.(g) *Every official act of the Commission shall be entered of record, and its hearings and records shall be open to the public. The Commission is authorized to make such rules as are necessary for the orderly transaction of its proceedings. Unless the Commission has adopted a dif-*

ferent rule, its proceedings shall be in accordance with the Federal Rules of Civil Procedure.

Sec. 12.(h) *The Commission may order testimony to be taken by deposition in any proceedings pending before it at any state of such proceeding. Any person may be compelled to appear and depose, and to produce books, papers, or documents, in the same manner as witnesses may be compelled to appear and testify and produce like documentary evidence before the Commission. Witnesses whose depositions are taken under this subsection, and the persons taking such depositions, shall be entitled to the same fees as are paid for like services in the courts of the United States.*

Sec. 12.(i) *For the purpose of any proceeding before the Commission, the provisions of Section 11 of the National Labor Relations Act (29 U.S.C. 161) are hereby made applicable to the jurisdiction and powers of the Commission.*

Sec. 12.(j) *A hearing examiner appointed by the Commission shall hear, and make a determination upon, any proceeding instituted before the Commission and any motion in connection therewith, assigned to such hearing examiner by the Chairman of the Commission, and shall make a report of any such determination which constitutes his final disposition of the proceedings. The report of the hearing examiner shall become the final order of the Commission within thirty days after such report by the hearing examiner, unless within such period any Commission member has directed that such report shall be reviewed by the Commission.*

Sec. 12.(k) *Except as otherwise provided in this Act, the hearing examiners shall be subject to the laws governing employees in the classified civil service, except that appointments shall be made without regard to Section 5108 of Title 5, United States Code. Each hearing examiner shall receive compensation at a rate not less than that prescribed for GS-16 under Section 5332 of Title 5, United States Code.*

STATE JURISDICTION AND STATE PLANS

Individual states have jurisdiction over safety and health issues when there is no federal OSHA standard in effect covering the issue.

Individual states may take over the OSHA programs in their own state under provisions of the act. Many states already have; check with the OSHA area director for approved state plan in effect in your state. (See the Directory at the back of this manual.)

"AS STATED"
Sec. 18. "THE ACT"

Sec. 18.(a) *Nothing in this Act shall prevent any State agency or court from asserting jurisdiction under State law over any occupational safety or health issue with respect to which no standard is in effect under Section 6.*

Sec. 18.(b) *Any State which, at any time, desires to assume responsibility for development and enforcement therein of occupational safety and health standards relating to any occupational safety or health issue with respect to which a Federal standard has been promulgated under Section 6 shall submit a State plan for the development of such standards and their enforcement.*

Sec. 18.(c) *The Secretary shall approve the plan submitted by a State under subsection (b), or any modification thereof, if such plan in his judgment—*

(1) designates a State agency or agencies as the agency or agencies responsible for administering the plan throughout the State,

(2) provides for the development and enforcement of safety and health standards relating to one or more safety or health issues, which standards (and the enforcement of which standards) are or will be at least as effective in providing safe and healthful employment and places of employment as the standards promulgated under Section 6

which relate to the same issues, and which standards, when applicable to products which are distributed or used in interstate commerce, are required by compelling local conditions and do not unduly burden interstate commerce,

(3) provides for a right of entry and inspection of all workplaces subject to the Act which is at least as effective as that provided in Section 8, and includes a prohibition on advance notice of inspections,

(4) contains satisfactory assurances that such agency or agencies have or will have the legal authority and qualified personnel necessary for the enforcement of such standards,

(5) gives satisfactory assurances that such State will devote adequate funds to the administration and enforcement of such standards,

(6) contains satisfactory assurances that such State will, to the extent permitted by its law, establish and maintain an effective and comprehensive occupational safety and health program applicable to all employees of public agencies of the State and its political subdivisions, which program is as effective as the standards contained in an approved plan,

(7) requires employers in the State to make reports to the Secretary in the same manner and to the same extent as if the plan were not in effect, and

(8) provides that the State agency will make such reports to the Secretary in such form and containing such information as the Secretary shall from time to time require.

Sec. 18.(d) *If the Secretary rejects a plan submitted under subsection (b), he shall afford the State submitting the plan due notice and opportunity for a hearing before so doing.*

Sec. 18.(e) *After the Secretary approves a State plan submitted under subsection (b), he may, but shall not be required to, exercise his authority under Sections 8, 9, 10, 13, and 17 with respect to comparable standards promulgated under Section 6, for the period specified in the next sentence. The Secretary may exercise the authority referred to above until he determines, on the basis of actual operations under the State plan, that the criteria set forth in subsection (c) are being applied, but he shall not make such determination for at least three years after the plan's approval under subsection (c). Upon making the determination referred to in the preceding sentence, the provisions of Sections 5 (a) (2), 8 (except for the purpose of carrying out subsection (f) of this section), 9, 10, 13, and 17, and standards promulgated under Section 6 of this Act, shall not apply with respect to any occupational safety or health issues covered under the plan, but the Secretary may retain jurisdiction under the above provisions in any proceeding commenced under Section 9 or 10 before the date of determination.*

Sec. 18.(f) *The Secretary shall, on the basis of reports submitted by the State agency and his own inspections, make a continuing evaluation of the manner in which each State having a plan approved under this section is carrying out such plan. Whenever the Secretary finds, after affording due notice and opportunity for a hearing, that in the administration of the State plan there is a failure to comply substantially with any provision of the State plan (or any assurance contained therein), he shall notify the State agency of his withdrawal of approval of such plan and upon receipt of such notice such plan shall cease to be in effect, but the State may retain jurisdiction in any case commenced before the withdrawal of the plan in order to enforce standards under the plan whenever the issues involved do not relate to the reasons for the withdrawal of the plan.*

Sec. 18.(g) *The State may obtain a review of a decision of the Secretary withdrawing approval of or rejecting its plan by the United States court of appeals for the circuit in which the State is located by filing in such court within thirty days following receipt of notice of such decision a petition to modify or set aside in whole or in part the action of the Secretary. A copy of such petition shall forthwith be served upon the Secretary, and thereupon the Secretary shall certify and file in the court the record upon which the decision complained of was issued as provided in Section 2112 of Title 28, United States Code. Unless the court finds that the Secretary's decision in rejecting a proposed State plan or withdrawing his approval of such a plan is not supported by substantial evidence, the court shall affirm the Secretary's decision. The judgment of the court shall be subject to review by the Supreme Court of the United States upon certiorari or certification as provided in Section 1254 of Title 28, United States Code.*

Sec. 18.(h) *The Secretary may enter into an agreement with a State under which the State will be permitted to continue to enforce one or more occupational health and safety standards in effect in such State until final action is taken by the Secretary with respect to a plan submitted by a State under subsection (b) of this section, or two years from the date of enactment of this Act, whichever is earlier.*

RESEARCH AND RELATED ACTIVITIES

A continual program for discovering and improving safety and health methods and problems is provided for as follows:

"AS STATED"
Sec. 20. "THE ACT"
Sec. 20.(a)
(1) The Secretary of Health, Education, and Welfare, after consultation with the Secretary and with other appropriate Federal departments or agencies, shall conduct (directly or by grants or contracts) research, experiments, and demonstrations relating to occupational safety and health, including studies of psychological factors involved, and relating to innovative methods, techniques, and approaches for dealing with occupational safety and health problems.

(2) The Secretary of Health, Education, and Welfare shall from time to time consult with the Secretary in order to develop specific plans for such research, demonstrations, and experiments as are necessary to produce criteria, in-

cluding criteria identifying toxic substances, enabling the Secretary to meet his responsibility for the formulation of safety and health standards under this Act; and the Secretary of Health, Education, and Welfare, on the basis of such research, demonstrations, and experiments and any other information available to him, shall develop and publish at least annually such criteria as will effectuate the purposes of this Act.

(3) The Secretary of Health, Education, and Welfare, on the basis of such research, demonstrations, and experiments, and any other information available to him, shall develop criteria dealing with toxic materials and harmful physical agents and substances which will describe exposure levels that are safe for various periods of employment, including but not limited to the exposure levels at which no employee will suffer impaired health or functional capacities or diminished life expectancy as a result of his work experience.

(4) The Secretary of Health, Education, and Welfare shall also conduct special research, experiments, and demonstrations relating to occupational safety and health as are necessary to explore new problems, including those created by new technology in occupational safety and health, which may require ameliorative action beyond that which is otherwise provided for in the operating provisions of this Act. The Secretary of Health, Education, and Welfare shall also conduct research into the motivational and behavioral factors relating to the field of occupational safety and health.

(5) The Secretary of Health, Education, and Welfare, in order to comply with his responsibilities under paragraph (2), and in order to develop needed information regarding potentially toxic substances or harmful physical agents, may prescribe regulations requiring employers to measure, record, and make reports on the exposure of employees to substances or physical agents which the Secretary of Health, Education, and Welfare reasonably believes may endanger the health or safety of employees. The Secre-

tary of Health, Education, and Welfare also is authorized to establish such programs of medical examinations and tests as may be necessary for determining the incidence of occupational illnesses and the susceptibility of employees to such illnesses. Nothing in this or any other provision of this Act shall be deemed to authorize or require medical examination, immunization, or treatment for those who object thereto on religious grounds, except where such is necessary for the protection of the health or safety of others. Upon the request of any employer who is required to measure and record exposure of employees to substances or physical agents as provided under this subsection, the Secretary of Health, Education, and Welfare shall furnish full financial or other assistance to such employer for the purpose of defraying any additional expense incurred by him in carrying out the measuring and recording as provided in this subsection.

(6) The Secretary of Health, Education, and Welfare shall publish within six months of enactment of this Act and thereafter as needed but at least annually a list of all known toxic substances by generic family or other useful grouping, and concentrations at which such toxicity is known to occur. He shall determine following a written request by any employer or authorized representative of employees, specifying with reasonable particularity the grounds on which the request is made, whether any substance normally found in the place of employment has potentially toxic effects in such concentrations as used or found; and shall submit such determination both to employers and affected employees as soon as possible. If the Secretary of Health, Education, and Welfare determines that any substance is potentially toxic at the concentrations in which it is used or found in a place of employment, and such substance is not covered by an occupational safety or health standard promulgated under Section 6, the Secretary of Health, Education, and Welfare shall immediately submit such determination to the Secretary, together with all pertinent criteria.

(7) Within two years of enactment of this Act, and an-

nually thereafter, the Secretary of Health, Education, and Welfare shall conduct and publish industrywide studies of the effect of chronic or low-level exposure to industrial materials, processes, and stresses on the potential for illness, disease, or loss of functional capacity in aging adults.

Sec. 20.(b) *The Secretary of Health, Education, and Welfare is authorized to make inspections and question employers and employees as provided in Section 8 of this Act in order to carry out his functions and responsibilities under this Section.*

Sec. 20.(c) *The Secretary is authorized to enter into contracts, agreements, or other arrangements with appropriate public agencies or private organizations for the purpose of conducting studies relating to his responsibilities under this Act. In carrying out his responsibilities under this subsection, the Secretary shall cooperate with the Secretary of Health, Education, and Welfare in order to avoid any duplication of efforts under this Section.*

Sec. 20.(d) *Information obtained by the Secretary and the Secretary of Health, Education, and Welfare under this section shall be disseminated by the Secretary to employers and employees and organizations thereof.*

Sec. 20.(e) *The functions of the Secretary of Health, Education, and Welfare under this Act shall, to the extent feasible, be delegated to the Director of the National Institute for Occupational Safety and Health established by Section 22 of this Act.*

TRAINING AND EMPLOYEE EDUCATION

Consultation and training provisions under OSHA are provided for as follows:

"AS STATED"
Sec. 21. "THE ACT"
Sec. 21.(a) *The Secretary of Health, Education, and Welfare, after consultation with the Secretary and with other*

appropriate Federal departments and agencies, shall conduct, directly or by grants or contracts

(1) education programs to provide an adequate supply of qualified personnel to carry out the purposes of this Act, and

(2) informational programs on the importance of and proper use of adequate safety and health equipment.

Sec. 21.(b) *The Secretary is also authorized to conduct, directly or by grants or contracts, short-term training of personnel engaged in work related to his responsibilities under this Act.*

Sec. 21.(c) *The Secretary, in consultation with the Secretary of Health, Education, and Welfare, shall*

(1) provide for the establishment and supervision of programs for the education and training of employers and employees in the recognition, avoidance, and prevention of unsafe or unhealthful working conditions in employments covered by this Act, and

(2) consult with and advise employers and employees, and organizations representing employers and employees, as to effective means of preventing occupational injuries and illnesses.

NATIONAL INSTITUTE FOR OCCUPATIONAL SAFETY AND HEALTH

The National Institute for Occupational Safety and Health (NIOSH) is provided to carry out the basis and purpose of the act and shall function as follows:

"AS STATED"

Sec. 22. "THE ACT"

Sec. 22.(a) *It is the purpose of this section to establish a National Institute for Occupational Safety and Health in the Department of Health, Education, and Welfare in order to carry out the policy set forth in Section 2 of this*

Act and to perform the functions of the Secretary of Health, Education, and Welfare under Sections 20 and 21 of this Act.

Sec. 22.(b) *There is hereby established in the Department of Health, Education, and Welfare a National Institute for Occupational Safety and Health. The Institute shall be headed by a Director who shall be appointed by the Secretary of Health, Education, and Welfare, and who shall serve for a term of six years unless previously removed by the Secretary of Health, Education, and Welfare.*

Sec. 22.(c) *The Institute is authorized to—*

(1) develop and establish recommended occupational safety and health standards; and

(2) perform all functions of the Secretary of Health, Education, and Welfare under Sections 20 and 21 of this Act.

Sec. 22.(d) *Upon his own initiative, or upon the request of the Secretary or the Secretary of Health, Education, and Welfare, the Director is authorized*

(1) to conduct such research and experimental programs as he determines are necessary for the development of criteria for new and improved occupational safety and health standards, and

(2) after consideration of the results of such research and experimental programs make recommendations concerning new or improved occupational safety and health standards. Any occupational safety and health standard recommended pursuant to this section shall immediately be forwarded to the Secretary of Labor, and to the Secretary of Health, Education, and Welfare.

Sec. 22.(e) *In addition to any authority vested in the Institute by other provisions of this section, the Director, in carrying out the functions of the Institute, is authorized to—*

(1) prescribe such regulations as he deems necessary governing the manner in which its functions shall be carried out;

(2) receive money and other property donated, bequeathed, or devised, without condition or restriction other than that it be used for the purposes of the Institute and to use, sell, or otherwise dispose of such property for the purpose of carrying out its functions;

(3) receive (and use, sell, or otherwise dispose of, in accordance with paragraph (2)), money and other property donated, bequeathed, or devised to the Institute with a condition or restriction, including a condition that the Institute use other funds of the Institute for the purposes of the gift;

(4) in accordance with the civil service laws, appoint and fix the compensation of such personnel as may be necessary to carry out the provisions of this section;

(5) obtain the services of experts and consultants in accordance with the provisions of Section 3109 of Title 5, United States Code;

(6) accept and utilize the services of voluntary and noncompensated personnel and reimburse them for travel expenses, including per diem, as authorized by Section 5703 of Title 5, United States Code;

(7) enter into contracts, grants, or other arrangements, or modifications thereof to carry out the provisions of this section, and such contracts or modifications thereof may be entered into without performance or other bonds, and without regard to Section 3709 of the Revised Statutes, as amended (41 U.S.C. 5), or any other provision of law relating to competitive bidding;

(8) make advance, progress, and other payments which the Director deems necessary under this title without regard to the provisions of Section 3648 of the Revised Statutes, as amended (31 U.S.C. 529); and

(9) make other necessary expenditures.

Sec. 22.(f) *The Director shall submit to the Secretary of Health, Education, and Welfare, to the President, and to the Congress an annual report of the operations of the Institute under this Act, which shall include a detailed statement of all private and public funds received and expended by it, and such recommendations as he deems appropriate.*

GRANTS TO STATES

The federal government will share in the cost of state OSHA programs as follows:

"AS STATED"

Sec. 23. "THE ACT"

Sec. 23.(a) *The Secretary is authorized, during the fiscal year ending June 30, 1971, and the two succeeding fiscal years, to make grants to the States which have designated a State agency under Section 18 to assist them—*

(1) in identifying their needs and responsibilities in the area of occupational safety and health,

(2) in developing State plans under Section 18, or

(3) in developing plans for—

(A) establishing systems for the collection of information concerning the nature and frequency of occupational injuries and diseases;

(B) increasing the expertise and enforcement capabilities of their personnel engaged in occupational safety and health programs; or

(C) otherwise improving the administration and enforcement of State occupational safety and health laws, including standards thereunder, consistent with the objectives of this Act.

Sec. 23.(b) *The Secretary is authorized, during the fiscal year ending June 30, 1971, and the two succeeding fiscal years, to make grants to the States for experimental and demonstration projects consistent with the objectives set forth in subsection (a) of this section.*

Sec. 23.(c) *The Governor of the State shall designate the appropriate State agency for receipt of any grant made by the Secretary under this section.*

Sec. 23.(d) *Any State agency designated by the Governor of the State desiring a grant under this section shall submit an application therefor to the Secretary.*

Sec. 23.(e) *The Secretary shall review the application, and shall, after consultation with the Secretary of Health, Education, and Welfare, approve or reject such application.*

Sec. 23.(f) *The Federal share for each State grant under subsection (a) or (b) of this section may not exceed 90 per centum of the total cost of the application. In the event the Federal share for all States under either such subsection is not the same, the differences among the States shall be established on the basis of objective criteria.*

Sec. 23.(g) *The Secretary is authorized to make grants to the States to assist them in administering and enforcing programs for occupational safety and health contained in State plans approved by the Secretary pursuant to Section 18 of this Act. The Federal share for each State grant under this subsection may not exceed 50 per centum of the total cost to the State of such a program. The last sentence of subsection (f) shall be applicable in determining the Federal share under this subsection.*

Sec. 23.(h) *Prior to June 30, 1973, the Secretary shall, after consultation with the Secretary of Health, Education, and Welfare, transmit a report to the President and to the Congress, describing the experience under the grant programs authorized by this section and making any recommendations he may deem appropriate.*

STATISTICS

The provisions for collecting, compilation, and analysis of statistics (recordkeeping) are as follows:

"AS STATED"
Sec. 24. "THE ACT"
Sec. 24.(a) *In order to further the purposes of this Act, the Secretary, in consultation with the Secretary of Health,*

Education, and Welfare, shall develop and maintain an effective program of collection, compilation, and analysis of occupational safety and health statistics. Such program may cover all employments whether or not subject to any other provisions of this Act but shall not cover employments excluded by Section 4 of the Act. The Secretary shall compile accurate statistics on work injuries and illnesses which shall include all disabling, serious, or significant injuries and illnesses, whether or not involving the loss of time from work, other than minor injuries requiring only first aid treatment and which do not involve medical treatment, loss of consciousness, restriction of work or motion, or transfer to another job.

Sec. 24.(b) *To carry out his duties under subsection (a) of this section, the Secretary may—*

(1) promote, encourage, or directly engage in programs of studies, information, and communication concerning occupational safety and health statistics;

(2) make grants to States or political subdivisions thereof in order to assist them in developing and administering programs dealing with occupational safety and health statistics; and

(3) arrange, through grants or contracts, for the conduct of such research and investigations as give promise of furthering the objectives of this section.

Sec. 24.(c) *The Federal share for each grant under subsection (b) of this section may be up to 50 per centum of the State's total cost.*

Sec. 24.(d) *The Secretary may, with the consent of any State or political subdivision thereof, accept and use the services, facilities, and employees of the agencies of such State or political subdivision, with or without reimbursement, in order to assist him in carrying out his functions under this section.*

Sec. 24.(e) *On the basis of the records made and kept pursuant to Section 8 (c) of this Act, employers shall file such reports with the Secretary as he shall prescribe by regula-*

tions, as necessary to carry out his functions under this Act.

Sec. 24.(f) *Agreements between the Department of Labor and States pertaining to the collection of occupational safety and health statistics already in effect on the effective date of this Act shall remain in effect until superseded by grants or contracts made under this Act.*

AUDITS

Those receiving grants under this act are subject to audits.

"AS STATED"
Sec. 25. "THE ACT"
Sec. 25.(a) *Each recipient of a grant under this Act shall keep such records as the Secretary or the Secretary of Health, Education, and Welfare shall prescribe, including records which fully disclose the amount and disposition by such recipient of the proceeeds of such grant, the total cost of the project or undertaking in connection with which such grant is made or used, and the amount of that portion of the cost of the project and undertaking supplied by other sources, and such other records as will facilitate an effective audit.*

Sec. 25.(b) *The Secretary or the Secretary of Health, Education, and Welfare, and the Comptroller General of the United States, or any of their duly authorized representatives, shall have access for the purpose of audit and examination to any books, documents, papers, and records of the recipients of any grant under this Act that are pertinent to any such grant.*

AGENCY'S ANNUAL REPORT

Each year a report on OSHA is prepared and submitted to the president for transmittal to Congress.

"AS STATED"

Sec. 26. "THE ACT"

Sec. 26. *Within one hundred and twenty days following the convening of each regular session of each Congress, the Secretary and the Secretary of Health, Education, and Welfare shall each prepare and submit to the President for transmittal to the Congress a report upon the subject matter of this Act, the progress toward achievement of the purpose of this Act, the needs and requirements in the field of occupational safety and health, and any other relevant information. Such reports shall include information regarding occupational safety and health standards, and criteria for such standards, developed during the preceding year; evaluation of standards and criteria previously developed under this Act, defining areas of emphasis for new criteria and standards; an evaluation of the degree of observance of applicable occupational safety and health standards, and a summary of inspection and enforcement activity undertaken; analysis and evaluation of research activities for which results have been obtained under governmental and nongovernmental sponsorship; an analysis of major occupational diseases; evaluation of available control and measurement technology for hazards for which standards or criteria have been developed during the preceding year; description of cooperative efforts undertaken between Government agencies and other interested parties in the implementation of this Act during the preceding year; a progress report on the development of an adequate supply of trained manpower in the field of occupational safety and health, including estimates of future needs and the efforts being made by Government and others to meet those needs; listing of all toxic substances in industrial usage for which labeling requirements, criteria, or standards have not yet been established; and such recommendations for additional legislation as are deemed necessary to protect the safety and health of the worker and improve the administration of this Act.*

NATIONAL COMMISSION ON STATE WORKMEN'S COMPENSATION LAWS

A National Commission on State Workmen's Compensation Laws has been created to study and evaluate state workmen's compensation laws to determine if such laws provide an adequate, prompt, and equitable system of compensation for injury or death arising out of or in the course of employment.

"AS STATED"
Sec. 27. "THE ACT"
Sec. 27.(a)(1) *The Congress hereby finds and declares that—*

(A) the vast majority of American workers, and their families, are dependent on workmen's compensation for their basic economic security in the event such workers suffer disabling injury or death in the course of their employment; and that the full protection of American workers from job-related injury or death requires an adequate, prompt, and equitable system of workmen's compensation as well as an effective program of occupational health and safety regulations; and

(B) in recent years serious questions have been raised concerning the fairness and adequacy of present workmen's compensation laws in the light of the growth of the economy, the changing nature of the labor force, increases in medical knowledge, changes in the hazards associated with various types of employment, new technology creating new risks to health and safety, and increases in the general level of wages and the cost of living.

Sec. 27.(a)(2) *The purpose of this section is to authorize an effective study and objective evaluation of State workmen's compensation laws in order to determine if such laws provide an adequate, prompt, and equitable system of compensation for injury or death arising out of or in the course of employment.*

Sec. 27.(b) *There is hereby established a National Commission on State Workmen's Compensation Laws.*

Sec. 27.(c)(1) *The Workmen's Compensation Commission shall be composed of fifteen members to be appointed by the President from among members of State workmen's compensation boards, representatives of insurance carriers, business, labor, members of the medical profession having experience in industrial medicine or in workmen's compensation cases, educators having special expertise in the field of workmen's compensation, and representatives of the general public. The Secretary, the Secretary of Commerce, and the Secretary of Health, Education, and Welfare shall be ex officio members of the Workmen's Compensation Commission.*

Sec. 27.(c)(2) *Any vacancy in the Workmen's Compensation Commission shall not affect its powers.*

Sec. 27.(c)(3) *The President shall designate one of the members to serve as Chairman and one to serve as Vice Chairman of the Workmen's Compensation Commission.*

Sec. 27.(c)(4) *Eight members of the Workmen's Compensation Commission shall constitute a quorum.*

Sec. 27.(d)(1) *The Workmen's Compensation Commission shall undertake a comprehensive study and evaluation of State workmen's compensation laws in order to determine if such laws provide an adequate, prompt, and equitable system of compensation. Such study and evaluation shall include, without being limited to, the following subjects:*

(A) the amount and duration of permanent and temporary disability benefits and the criteria for determining the maximum limitations thereon,

(B) the amount and duration of medical benefits and provisions insuring adequate medical care and free choice of physician,

(C) the extent of coverage of workers, including exemptions based on numbers or type of employment,

(D) standards for determining which injuries or diseases should be deemed compensable,

(E) rehabilitation,

(F) coverage under second or subsequent injury funds,

(G) time limits on filing claims,

(H) waiting periods,

(I) compulsory or elective coverage,

(J) administration,

(K) legal expenses,

(L) the feasibility and desirability of a uniform system of reporting information concerning job-related injuries and diseases and the operation of workmen's compensation laws,

(M) the resolution of conflict of laws, extraterritoriality, and similar problems arising from claims with multistate aspects,

(N) the extent to which private insurance carriers are excluded from supplying workmen's compensation coverage and the desirability of such exclusionary practices, to the extent they are found to exist,

(O) the relationship between workmen's compensation on the one hand, and old-age, disability, and survivors insurance and other types of insurance, public or private, on the other hand,

(P) methods of implementing the recommendations of the Commission.

Sec. 27.(d)(2) *The Workmen's Compensation Commission shall transmit to the President and to the Congress not later than July 31, 1972, a final report containing a detailed statement of the findings and conclusions of the Commission, together with such recommendations as it deems advisable.*

Sec. 27.(e)(1) *The Workmen's Compensation Commission or, on the authorization of the Workmen's Compensa-*

tion Commission, any subcommittee or members thereof may, for the purpose of carrying out the provisions of this title, hold such hearings, take such testimony, and sit and act at such times and places as the Workmen's Compensation Commission deems advisable. Any member authorized by the Workmen's Compensation Commission may administer oaths or affirmations to witnesses appearing before the Workmen's Compensation Commission or any subcommittee or members thereof.

Sec. 27.(e)(2) *Each department, agency, and instrumentality of the executive branch of the Government, including independent agencies, is authorized and directed to furnish to the Workmen's Compensation Commission, upon request made by the Chairman or Vice Chairman, such information as the Workmen's Compensation Commission deems necessary to carry out its functions under this section.*

Sec. 27.(f) *Subject to such rules and regulations as may be adopted by the Workmen's Compensation Commission, the Chairman shall have the power to—*

(1) appoint and fix the compensation of an executive director, and such additional staff personnel as he deems necessary, without regard to the provisions of Title 5, United States Code, governing appointments in the competitive service, and without regard to the provisions of Chapter 51 and subchapter III of Chapter 53 of such title relating to classification and General Schedule pay rates, but at rates not in excess of the maximum rate for GS-18 of the General Schedule under Section 5332 of such title, and

(2) procure temporary and intermittent services to the same extent as is authorized by Section 3109 of Title 5, United States Code.

Sec. 27.(g) *The Workmen's Compensation Commission is authorized to enter into contracts with Federal or State agencies, private firms, institutions, and individuals for the conduct of research or surveys, the preparation of reports, and other activities necessary to the discharge of its duties.*

Sec. 27.(h) *Members of the Workmen's Compensation Commission shall receive compensation for each day they are engaged in the performance of their duties as members of the Workmen's Compensation Commission at the daily rate prescribed for GS-18 under Section 5332 of Title 5, United States Code, and shall be entitled to reimbursement for travel, subsistence, and other necessary expenses incurred by them in the performance of their duties as members of the Workmen's Compensation Commission.*

Sec. 27.(i) *There are hereby authorized to be appropriated such sums as may be necessary to carry out the provisions of this section.*

Sec. 27.(j) *On the ninetieth day after the date of submission of its final report to the President, the Workmen's Compensation Commission shall cease to exist.*

ECONOMIC ASSISTANCE TO SMALL BUSINESSES

Small business loans to fulfill compliance with OSHA are provided for.

"AS STATED"
Sec. 28. "THE ACT"
Sec. 28.(a) *Section 7(b) of the Small Business Act, as amended, is amended—*

(1) by striking out the period at the end of "paragraph (5)" and inserting in lieu thereof ";and"; and

(2) by adding after paragraph (5) a new paragraph as follows:

"(6) to make such loans (either directly or in cooperation with banks or other lending institutions through agreements to participate on an immediate or deferred basis) as the Administration may determine to be necessary or appropriate to assist any small business concern in effect-

ing additions to or alterations in the equipment, facilities, or methods of operation of such business in order to comply with the applicable standards promulgated pursuant to Section 6 of the Occupational Safety and Health Act of 1970 or standards adopted by a State pursuant to a plan approved under Section 18 of the Occupational Safety and Health Act of 1970, if the Administration determines that such concern is likely to suffer substantial economic injury without assistance under this paragraph."

Sec. 28.(b) *The third sentence of Section 7(b) of the Small Business Act, as amended, is amended by striking out "or (5)" after "paragraph (3)" and inserting a comma followed by "(5) or (6)."*

Sec. 28. (c) *Section 4 (c) (1) of the Small Business Act, as amended, is amended by inserting "7(b) (6)," after "7(b) (5),".*

Sec. 28.(d) *Loans may also be made or guaranteed for the purposes set forth in Section 7(b) (6) of the Small Business Act, as amended, pursuant to the provisions of Section 202 of the Public Works and Economic Development Act of 1965, as amended.*

ADDITIONAL ASSISTANT SECRETARY OF LABOR

Section 2 of the Act of April 17, 1946 is amended:

"AS STATED"
Sec. 29. "THE ACT"
Sec. 29.(a) *Section 2 of the Act of April 17, 1946 (60 Stat. 91) as amended (29 U.S.C. 553) is amended by—*

(1) striking out "four" in the first sentence of such section and inserting in lieu thereof "five"; and

(2) adding at the end thereof the following new sentence: "One of such Assistant Secretaries shall be an Assistant Secretary of Labor for Occupational Safety and Health."

Sec. 29.(b) *Paragraph (20) of Section 5315 of Title 5, United States Code, is amended by striking out "(4)" and inserting in lieu thereof "(5)."*

ADDITIONAL POSITIONS

Section 5108(c) of Title 5, United States Code, is amended:

"AS STATED"
Sec. 30. "THE ACT"
Sec. 30. *Section 5108 (c) of Title 5, United States Code, is amended by—*

(1) striking out the word "and" at the end of paragraph (8);

(2) striking out the period at the end of paragraph (9) and inserting in lieu thereof a semicolon and the word "and"; and

(3) by adding immediately after paragraph (9) the following new paragraph:

"(10) (A) the Secretary of Labor, subject to the standards and procedures prescribed by this chapter, may place an additional twenty-five positions in the Department of Labor in GS-16, 17, and 18 for the purposes of carrying out his responsibilities under the Occupational Safety and Health Act of 1970;

"(B) the Occupational Safety and Health Review Commission, subject to the standards and procedures prescribed by this chapter, may place ten positions in GS-16, 17, and 18 in carrying out its functions under the Occupational Safety and Health Act of 1970."

EMERGENCY LOCATOR BEACONS

Section 601 of the Federal Aviation Act of 1958 is amended:

"AS STATED"
Sec. 31. "THE ACT"

Sec. 31. *Section 601 of the Federal Aviation Act of 1958 is amended by inserting at the end thereof a new subsection as follows:*

"EMERGENCY LOCATOR BEACONS"

"(d) (1) Except with respect to aircraft described in paragraph (2) of this subsection, minimum standards pursuant to this section shall include a requirement that emergency locator beacons shall be installed—

"(A) on any fixed-wing, powered aircraft for use in air commerce the manufacture of which is completed, or which is imported into the United States, after one year following the date of enactment of this subsection; and

"(B) on any fixed-wing, powered aircraft used in air commerce after three years following such date.

"(d) (2) The provisions of this subsection shall not apply to jet-powered aircraft; aircraft used in air transportation (other than air taxis and charter aircraft); military aircraft; aircraft used solely for training purposes not involving flights more than twenty miles from its base; and aircraft used for the aerial application of chemicals."

SEPARABILITY

Invalid provisions of the act.

"AS STATED"
Sec. 32. "THE ACT"
Sec. 32. *If any provision of this Act, or the application of such provision to any person or circumstance, shall be held invalid, the remainder of this Act, or the application of such provision to persons or circumstances other than those as to which it is held invalid, shall not be affected thereby.*

APPROPRIATIONS

Authorized appropriations:

"AS STATED"
Sec. 33. "THE ACT"
Sec. 33. *There are authorized to be appropriated to carry out this Act for each fiscal year such sums as the Congress shall deem necessary.*

EFFECTIVE DATE

Effective date, April 28, 1971.

"AS STATED"
Sec. 34. "THE ACT"
Sec. 34. *This Act shall take effect one hundred and twenty days after the date of its enactment. Approved December 29, 1970.*

LEGISLATIVE HISTORY

House reports: *No. 91-1291 accompanying H.R. 16785 (Comm. on Education and Labor) and No. 91-1765 (Comm. of Conference).*

Senate report: *No. 91-1282 (Comm. on Labor and Public Welfare).*

Congressional Record, Vol. 116 (1970):
Oct. 13, Nov. 16, 17, considered and passed Senate.
Nov. 23, 24, considered and passed House, amended, in lieu of H.R. 16785.
Dec. 16, Senate agreed to conference report.
Dec. 17, House agreed to conference report.

15

AN EFFECTIVE ACCIDENT PREVENTION SAFETY PROGRAM

AN EFFECTIVE ACCIDENT PREVENTION SAFETY PROGRAM

Of the three basic areas of concern in OSHA compliance, employer–employee participation can be the most important.

The first stated goal of the OSHA law is to reduce hazards and to institute and perfect existing programs for providing safe and healthful conditions.

> "AS STATED"
> Sec. 2. "THE ACT"
> Sec. 2.(b)(1) *by encouraging employers and employees in their efforts to reduce the number of occupational safety and health hazards at their places of employment, and to stimulate employers and employees to institute new and to perfect existing programs for providing safe and healthful working conditions;*

Aside from complying with the law, which you must, think of the tragedy of anyone being killed or needlessly injured. Furthermore, think of the aftereffect and potential liabilities.

Let us suppose for a moment that you are called to testify in court on your own behalf or on behalf of your employer in a case where someone at your workplace has been maimed or killed, and the dependents or survivors are suing for negligence. Three basic questions are asked:

1 / Please explain to the court the safety and health laws that apply to your workplace.
2 / Please explain to the court what programs exist to ensure safe and healthful working conditions for all employees who work there.
3 / Have all employees been informed of such programs and, if so, how?

Can you or could you answer any one or all of these questions sufficiently to prove that there was no negligence in any of these three areas at your workplace? This case obviously cites an extreme; however, it could be a very real experience for you.

The old adage that "an ounce of prevention is worth a pound of cure" still holds true, and especially in safety and health matters.

AN EFFECTIVE WAY TO ASSURE COMPLIANCE AND FULFILL YOUR RESPONSIBILITIES

You have already taken a big step in becoming informed of the requirements of this mandatory law by reading and reviewing this manual. Now make sure that each of your employees or fellow employees are as informed as you. Have each of them read this manual. You may possibly want to require them to have their own copy, or you may even wish to provide them with a copy. Make sure that *all* are informed! Ignorance of the law is no excuse.

Once the mandatory requirements are understood, it is time to put into action an effective program for carrying out these requirements. The most positive way of accomplishing this is to establish your own safety program. There are no designated criteria for the formulation or structuring of such a program; however, the following suggestions are offered:

1 / Keep it as simple as possible.

2 / Keep it clear.

3 / Make sure it is thorough.

4 / Make sure it is effective.

WHAT SHOULD AN ACCIDENT PREVENTION SAFETY PROGRAM CONTAIN?

Safety programs can be structured in many different ways. They should be carefully structured to cover the needs of each area of work.

The following six items offer a basis on which to start to build such a program:

1 / Title of program: For the purpose of discussion in this manual, we have chosen to use the title Accident Prevention Safety Program. Other appropriate titles can be used, such as Loss Prevention Program, Safety Program, etc.

2 / Certification of adoption: Once your accident prevention safety program has been finalized, you will want to present it to all involved. You should clearly establish that this is your program covering all work areas. This can easily be done by the inclusion of a certification of adoption properly executed. See sample included herein.

3 / Statement of safety and health policies: This is your statement of policies in writing so there can be no question of your position. The sample included herein has been prepared on a broad application basis. You may wish to add to or modify it to your own individual requirements.

4 / Safety practices and operations code: This code establishes the do's and don'ts directly related to safety. Again, the sample included herein has been prepared on a broad application basis. You may wish to add or modify it to your own individual requirements.

5 / Safety meetings: Safety meetings are an important part of any program. The sample included herein is an organized program for holding required safety meetings, again in writing so that everyone knows what to do.

6 / Job inspection and accident investigation: Establish procedures to keep workplaces safe, and through accident investigation determine the cause of the accident to help prevent future injuries.

The above six items form a basis. Now it is up to you. If you follow these suggestions, and add some of your own, you should end up with a pretty good program. Get into the habit of equating productivity with "protectivity." A safer plant is a better place to work.

To structure a single accident prevention safety program to cover all workers on each and every occasion would be nearly impossible. However, the following accident prevention safety program, as presented, should apply to most applications and will provide helpful information for others.

1—TITLE

ACCIDENT PREVENTION
SAFETY PROGRAM

2—CERTIFICATION OF ADOPTION

ACCIDENT PREVENTION SAFETY PROGRAM

The following accident prevention program is hereby adopted for all employment activities executed on behalf of or at the direction or request of the following:

Company Name: ____________________

Address: ____________________

City or Location: ____________________

Phone Number: ____________________

Responsible Party's Signature: ____________________

Adopted This Date: ____________________

Note: Attention shall be given from time to time to changes in current state and federal requirements so that compliance may be followed.

3—STATEMENT OF SAFETY AND HEALTH POLICIES

Safety and Health Precedence

The personal and collective safety and health of all employees is of primary importance. The prevention of occupationally (work-related) caused injuries and illnesses is of such consequence that it shall be given precedence over operating productivity.

Safety Practices

Safety shall be practiced by all personnel at all times. Only safe methods and equipment shall be used.

Cooperation

Effective standards for guarding against injuries and illnesses while on the job shall be continuously maintained. To be successful, proper attitudes toward the prevention of injuries and illness on the part of all employees are required. Success in all safety and health matters also depends upon cooperation among the company, its supervisors, and all employees, and also between each employee and fellow workers. Only through such cooperative attitudes and efforts can a safety record in the best interest of all be established and preserved.

Goal

This safety and health program is designed to reduce the number of injuries and illnesses to a minimum. *The goal is for zero accidents, injuries, and illnesses.*

Safety Standards

The safety standards include:

- Complying fully and completely with safety laws, rules, and regulations.
- Requiring that *all* employees comply and cooperate with all safety and health rules.
- Conducting safety meetings to provide education and training on safety and health matters.
- Conducting safety and health inspections to identify and eliminate unsafe working conditions and/or practices.
- Investigating, promptly and thoroughly, every accident to determine what caused it and to correct the problem so that it will not happen again.

Additional Provisions

Recognizing that the responsibilities for safety and health must be shared by all, the following is further established:

The responsibility for enacting, maintaining, and improving safety and health standards is the responsibility of each individual and requires full cooperation toward the prevention of occupational (job-related) accidents, injuries, and/or illnesses by every individual.

Supervisors shall be responsible for developing the proper attitudes toward safety and health in themselves and in those they supervise, and for ensuring that all operations are performed with the utmost regard for the safety and health of all personnel involved, including themselves.

Supervisors shall be responsible for seeing that all employees are properly trained in the safe performance of their job.

Supervisors shall be responsible for documenting properly and thoroughly all matters relating to safety and health.

All employees shall be responsible for cooperating wholeheartedly, with all aspects of safety and health, including complying with all rules and regulations, and for continuously practicing safety while performing their duties.

4—SAFETY PRACTICES AND OPERATIONS CODE

- All employees shall follow safe practices, use personal protective equipment as required, render every possible aid to safe operations, and report all unsafe conditions or practices.
- Work shall be well planned and supervised to prevent injuries.
- All employees shall be given frequent accident prevention instructions.
- Supervisors shall insist on employees observing and obeying every rule, regulation, and order necessary for the safe conduct of the work.
- All unsafe, unhealthy, or hazardous conditions or places shall be immediately placed off limits, out of order, out of bounds, etc., and then promptly removed or corrected.
- No one shall knowingly be permitted or required to work with impaired ability or alertness caused by fatigue, illness, or other factors such that the employee or others may be exposed to accidents or injury.
- No one will be allowed on the job while under the influence of intoxicating liquor or drugs.
- Horseplay, scuffling, and other acts that have or tend to have an adverse influence on the safety or well-being of employees are prohibited.
- Crowding or pushing when boarding or leaving any vehicle or other conveyance is prohibited.
- Employees shall be alert to see that all guards and other protective devices are in their proper places and adjusted prior to operating equipment and shall report deficiencies promptly.
- Workers shall not handle or tamper with any tools, equipment, machinery, or facilities not within the scope of their duties, unless they are thoroughly qualified and have received instructions from their supervisor.
- All injuries shall be reported promptly, so that arrangements can be made for medical or first aid treatment.

- When lifting heavy objects, use the large muscles of the legs instead of the smaller muscles of the back.
- Protect the eyes at all times through the proper use of goggles, hoods, etc.
- Know where you are going and how you are going to get there. Look before you move.
- Watch out for others; they may not be aware of what you are doing or where you are going.
- Wash thoroughly after handling injurious or poisonous substances, and follow all special instructions from authorized sources. Hands should be thoroughly cleaned just prior to eating.
- Loose or frayed clothing, dangling ties, finger rings, etc., shall not be worn near moving machinery or other sources of entanglement.
- Apparatus, tools, equipment, and machinery shall not be repaired or adjusted while in operation, nor shall oiling of moving parts be attempted, except on equipment that is designed or fitted with safeguards to protect the person performing the work.
- Use common sense. If you do not know, don't do it.

Additional helpful reminders for preventing accidents, injuries, and illnesses may be added to the above.

5—SAFETY MEETINGS

All employees shall be kept constantly aware of safe work methods. Safety meetings shall be well planned and held on a regular schedule as follows:

Field, Shop, Warehouse, and Yard Crews

A five- to ten-minute safety (tool box) meeting shall be held weekly for each crew.

All Other Company Personnel

Others shall attend safety meetings as directed. Such meetings will be scheduled so that safe work requirements for your operation will be covered.

Meeting Leaders

Supervisors are responsible for holding safety meetings. One subject only should be chosen and presented by the leader for each meeting.

Subject Matter

Safety meetings are not visiting sessions. Literature and information is available from the state and the National Safety Council, 525 N. Michigan Avenue, Chicago, Illinois 60611, for such subjects as "How to Lift," "How to Keep from Falling," "Eye Protection," etc. However, supervisors should use their own experiences and observations in making the presentations. Supervisors should develop safety meeting subjects and material based on a survey of the needs of the jobs being covered rather than using extemporaneous subjects.

Safety Meeting Notice

All employees shall be notified of the safety meetings that they are required to attend. It is recommended that a notice be posted well in advance of the meeting (see sample copy) at a place where all personnel who are required to attend will see it.

Safety Meeting Record

All safety meetings shall be recorded as follows: A safety meeting report (see sample copy) shall be completed for each and every safety meeting held. Immediately following the meeting, the report shall be transmitted to the company office where it shall be maintained in the company's safety files.

The record shall note the following:

Project name: Note the name of the project and/or group for which the meeting is being held.

Date and time: The date and approximate time of day or night the safety meeting was held.

Meeting leader: The name of the supervisor holding the meeting.

Employees in attendance: The name *printed clearly* and the signature of all employees in attendance, including the meeting leader.

Subject and comments: A brief description of the subject and/or subjects covered in the meeting with any comments.

Remember: Compliance and cooperation with all safety and health rules and regulations is a requirement for everyone.

Safety meetings are a *requirement*. Conduct and/or attend them as though your life depended on them.

SAFETY MEETING NOTICE

NOTICE TO EMPLOYEES

Crew or Group ____________

A Safety Meeting will be held . . .

WHERE: ____

WHEN: ____

MEETING LEADER: ________________________________

Please be prompt . . . The meeting will be brief . . . Attendance is mandatory.

SAFETY MEETING REPORT

PROJECT NAME: ______________________

DATE: __________ APPROX. TIME: ________

DURATION: ____________

MEETING LEADER: ______________________

SUBJECT AND COMMENTS:				PERSONNEL IN ATTENDANCE	
				EMPLOYEE'S NAME (Print Clearly)	EMPLOYEE'S SIGNATURE

Reduced Size Facsimile

6—JOB INSPECTION AND ACCIDENT INVESTIGATION

Job Inspection

A thorough inspection must be made of each jobsite and/or workplace or work station at least once each week. Even areas that are used infrequently must be inspected weekly.

Housekeeping, stairway lighting, work methods, equipment, tools, etc., should be examined to discover hazardous conditions that may have escaped routine detection. Special attention should be paid to things or conditions that have caused injury or illness on the particular job or on other jobs.

Any and all unsafe or unhealthy places or conditions shall be immediately corrected or reported to the nearest company supervisor. A safety and health inspection and report record (see sample) shall be used to record *all* unsafe and/or unhealthy conditions. This record shall be kept current and maintained in a clean, clear, and legible manner.

Any and all activities are subject to immediate inspection at any time. All activities shall be conducted in full compliance with applicable safety standards and shall be ready for a compliance inspection at all times.

Accident Investigation

Each accident must be investigated immediately by a company supervisor.

Corrective action shall be taken promptly. *Remember:* by correcting past mistakes we can help prevent future injuries.

Every occupational (work-related) injury or occupational disease that (1) results in lost time beyond the day of injury, or (2) requires medical treatment other than first aid must be reported in writing by the supervisor and sent to the company office within twenty-four hours of the accident.

In addition, if the injury results in a death or probable death, or hospitalization of two or more employees, the nearest OSHA office must be notified immediately.

SAFETY AND HEALTH INSPECTION AND REPORT RECORD

COMPANY NAME ________________ PROJECT NAME ________________

ITEM (Description of Condition Found)	AREA	DATE DETECTED	RECOMMENDATIONS AND ACTION TAKEN	TARGET DATE FOR CORRECTION	DATE CORRECTED

Reduced Size Facsimile

16

BRINGING YOUR WORKPLACE INTO COMPLIANCE

BRINGING YOUR WORKPLACE INTO COMPLIANCE

Can Your Workplace Pass an OSHA Inspection?

Since OSHA went into effect in 1971, less than 23 per cent of all firms inspected have been found to be in compliance.

Who Is Responsible?

Under OSHA it is the employer who has the obligation and responsibility to provide safe and healthful workplaces. (See Chapter 3, Section 5, "The Act.") There are numerous other requirements of both the employer and the employee (see Table of Contents); however, to ensure safe and healthful working conditions for all is the stated goal of OSHA.

Controlling the Working Conditions

Virtually everyone must comply with OSHA; however, it is the employer who has the direct responsibility to control the conditions in the workplace. Without proper control, hazardous conditions and injuries can result. Furthermore, if hazardous conditions are found by an OSHA inspector, fines will be levied against the employer. (See Chapter 5.)

> • **EXAMPLE** • The presence of a rotted and unsafe guard rail at a height from which a fall would probably result in an injury would demonstrate a lack of control of the conditions in the area of the guard rail.

Thorough and frequent inspections are required to control working conditions. Chapter 15 of this manual includes a basic procedure for maintaining job inspections and accident investigation.

Your program for prevention and control

Chapter 15 of this manual is a basic "Accident Prevention Safety Program" designed to help prevent accidents before they happen, as well as provide for investigations after each accident in order to prevent reoccurrence.

Your program for prevention and control (detecting and eliminating) of both immediate and basic causes of injuries and illnesses is the best means for control of safety and health matters in the workplace.

Safety and Health Standards

The basic minimum standards which you must meet are set by OSHA. (See Chapter 12 of this manual.)

- Identify those standards which apply to your operation and get a copy immediately.
- Check your operation against the standards to determine compliance.
- Develop and implement a plan to bring your operation into compliance.
- Thoroughly review each employee's job to determine what hazards are involved and how the hazards may be eliminated or controlled.
- Inspect regularly on a continuing basis.
- Make a checklist so that you do not overlook any condition or act which might result in an injury or illness.
- Make sure that proper maintenance is regularly performed.
- Make sure that everyone is properly trained to perform his/her job safely.

Your Checklist

As previously mentioned, you should make a checklist so that you do not and can not overlook any condition or act which might result in an injury or illness.

The following checklist will help you get started:

HOUSEKEEPING	OSHA STANDARD	CONDITION
Workplaces clean & orderly	1910.22(a)(1)	
Floors clean & dry	1910.22(a)(2)	
Floors in good repair	1910.22(a)(3)	
Aisles & passageways clear	1910.22(b)(1)	
Aisles & passageways marked	1910.22(b)(2)	
Covers & guardrails in place	1910.22(c)	
Floor loading plates in place	1910.22(d)(1)	
Safe floor & roof loading	1910.22(d)(2)	

MEANS OF EGRESS	OSHA STANDARD	CONDITION
Adequate emergency exits	1910.36(b)(1)	
Exits clear & unlocked	1910.36(b)(4)	
Exits clearly marked	1910.36(b)(5)	
Exits lighted	1910.36(b)(6)	

STAIRWAYS	OSHA STANDARD	CONDITION
Stair railings	1910.23(d)	
Stair railings—OSHA specifications	1910.23(e)	
Stair strength	1910.24(c)	
Stair steepness	1910.24(e)	
Stair length	1910.24(g)	
Tread gripping	1910.24(f)	
Overhead clearance	1910.24(i)	

FIRE PROTECTION	OSHA STANDARD	CONDITION
Fire extinguishers charged	1910.157(a)(1)	
Number of fire extinguishers	1910.157(b)	
Extinguisher locations	1910.157(a)(2)	
Location marking of extinguishers	1910.157(a)(3)	
Extinguisher mountings	1910.157(a)(5)	
Special purpose extinguishers	1910.157(a)(4)	
Automatic sprinklers	1910.159(a)(1)	
Water supply	1910.159(a)(2)	
Fire department connection	1910.159(b)	
Sprinkler system maintenance	1910.36(d)(1)	
Fire alarm boxes	1910.163(b)	
Fire alarm testing	1910.37(n)	

EMPLOYEE FACILITIES	OSHA STANDARD	CONDITION
Drinking & washing water available	1910.141(b)(1)	
Men & women's toilets	1910.141(c)(1)	
Toilet room construction	1910.141(c)(2)	

MEDICAL & FIRST AID	OSHA STANDARD	CONDITION
Hospital or medical facility	1910.151(b)	______
First aid trained personnel	1910.151(b)	______
Medical personnel available	1910.151(a)	______
Eye-wash emergency facilities	1910.151(b)	______

STORAGE OF MATERIALS	OSHA STANDARD	CONDITION
Safe clearances	1910.176(a)	______
Stacked materials	1910.176(b)	______
Storage areas kept free	1910.176(c)	______
Drainage system	1910.176(d)	______
Warning signs	1910.176(e)	______

MACHINE GUARDING	OSHA STANDARD	CONDITION
Are machines guarded	1910.212(a)(1)	______
Additional hazards	1910.212(a)(2)	______
Guard locations	1910.212(a)(3)	______
Machinery anchored	1910.212(b)	______

FLOOR & WALL OPENINGS	OSHA STANDARD	CONDITION
Floor holes guarded	1910.23(a)(8)	______
Pits & trapdoors guarded	1910.23(a)(5)	______
Toe boards on ladderway or platforms	1910.23(a)(2)	______
Skylights protection	1910.23(a)(4)	______
Temporary floors protection	1910.23(a)(7)	______
Wall openings guarded	1910.23(b)	______
Open-sided floors, platforms	1910.23(c)	______

PERSONAL PROTECTIVE EQUIPMENT	OSHA STANDARD	CONDITION
Protective equipment used & maintained	1910.132(a)	______
Reliability ratings	1910.132(a)	______
Employee-provided equipment	1910.132(b)	______
Eye protection used	1910.133(a)	______
Eye protection construction	1910.133(a)(2)	______
Respirators provided & used	1910.134(a)	______
Respirator-use training	1910.134(e)(5)	______
Respirator check	1910.134(f)(1)	______
Respirator cleanliness	1910.134(f)(3)	______
Respirators meeting OSHA specifications	1910.134(c)	______
Hard hats: use & construction	1910.135	______
Safety shoes	1910.136	______

The above checklist is provided only to help get you started. Make your own—one particularly applicable to all your workplaces and use it in your regular inspections.

Record Each Hazard Found

Use of the "Safety and Health Inspection and Report Record" form provided in Chapter 15 will give you a record that details any noncompliance.

Health Hazards

Health Standards are covered in OSHA Standards Volume #1—Part 1910—Subpart G—"Occupational Health and Environment Control" (See Chapter 12). It is recommended that you review these standards thoroughly to eliminate and guard against health hazards. The detecting of health hazards can pose special problems since they may not always be easy to detect without special detection equipment and should be dealt with carefully.

OSHA Standards Part 1910.94 deals with Ventilation
OSHA Standards Part 1910.95 deals with Occupational Noise Exposure
OSHA Standards Part 1910.96 & 1910.97 deals with Radiation

It is recommended that you consult experts in the related field to detect and eliminate these special hazards.

Priorities for Correction

Once any such hazards are found, you should set correction priorities. By setting priorities, the most serious condition can be corrected first, the next most serious next, and so on.

1 / A condition or practice where there is substantial probability that the consequence of an accident resulting from the condition would be death or serious physical harm.

2 / A condition or practice where an accident or illness resulting from the condition would probably not cause death or serious physical harm, but would have a direct or immediate effect on the safety or health of employees.

Determine Corrective Measures

Carefully analyze each hazardous condition found. Break down the work area and the tasks performed in it into basic parts. By so doing, you will be able to obtain an over-all view leading to a safer environment rather than possibly to additional hazards created by means of simple stop-gap methods. Call in safety specialists; many times they can save you time, effort, and money in helping you solve your problems.

Now Correct the Condition and Eliminate the Hazards

Through intelligent investigation and planning, you stand a much better chance of eliminating hazards and coming into compliance before an accident occurs or before the OSHA inspector arrives.

Use All Your Resources

Do not make the mistake of thinking you know it all. Consult with others around you for assistance in locating and eliminating hazards.

- Consult with employees; their observations and evaluations can be helpful.
- Contact employers' associations.
- Check with unions representing your workers.
- Consult with your insurance carrier. Most insurance companies provide invaluable help as a part of their service to you.
- Call in safety experts to help locate and solve problems before they happen.
- Subscribe to safety publications.
- Try a NIOSH walk-through (see Chapter 1).

Keys to Safety and Health Control

The following are important to the control of safety and health in your workplaces:

- Your commitment to bring your workplaces into compliance and provide safe and healthful workplaces.
- Your program for coming into and maintaining continual compliance.
- On-going inspections and corrective measures.
- Involve your supervisors. They can be your continual safety inspectors.
- Create and maintain an attitude and atmosphere of safety and health in all your workplaces.

Remember: Your attitude and cooperation towards safety and health compliance is important (see Chapter 4). When the OSHA inspector arrives, show him your programs for coming into and maintaining compliance. If you have done your job well, you will without question get a more favorable evaluation.

DIRECTORY FOR OSHA OFFICES

DIRECTORY FOR OSHA OFFICES

It should be noted that certain OSHA offices may have moved while this *Manual* was in preparation. Users are advised to verify area or field addresses by telephoning or writing the appropriate regional office.

REGION I: *CONNECTICUT, MAINE, MASSACHUSETTS, NEW HAMPSHIRE, RHODE ISLAND, VERMONT*

Regional Office
U.S. Department of Labor—OSHA
18 Oliver Street
Boston, Massachusetts 02110
Phone: area code 617—223-6712

Boston Area Office
U.S. Department of Labor—OSHA
Custom House Building, Room 703
State Street
Boston, Massachusetts 02109
Phone: area code 617—223-4511

Concord Area Office
U.S. Department of Labor—OSHA
Federal Building—Room 426
55 Pleasant Street
Concord, New Hampshire 03301
Phone: area code 603—224-1995

Hartford Area Office
U.S. Department of Labor—OSHA
Federal Building—Room 617B
450 Main Street
Hartford, Connecticut 06103
Phone: area code 203—244-2294

Providence District Office
U.S. Department of Labor—OSHA
Federal Building—Room 503A
U.S. Courthouse
Providence, Rhode Island 02903
Phone: area code 401—528-4466

Springfield Area Office
U.S. Department of Labor—OSHA
U.S. Post Office and Courthouse
436 Dwight Street
Springfield, Massachusetts 01103
Phone: area code 413—781-2420, extension 312

REGION II: *NEW JERSEY, NEW YORK, CANAL ZONE, PUERTO RICO, VIRGIN ISLANDS*

Regional Office
U.S. Department of Labor—OSHA
1515 Broadway (1 Astor Plaza)—Room 3445
New York, New York 10036
Phone: area code 212—971-5941

Albany Field Operations Office
U.S. Department of Labor—OSHA
112 State Street—Room 1120
Albany, New York 12207
Phone: area code 518—472-6085

Buffalo Field Operations Office
U.S. Department of Labor—OSHA
111 West Huron Street—Room 1002
Buffalo, New York 14202
Phone: area code 716—842-3333

Garden City Field Operations Office
U.S. Department of Labor—OSHA
370 Old Country Road
Garden City, New York 11530
Phone: area code 516—294-0400

Long Island Area Office
U.S. Department of Labor—OSHA
370 Old Country Road
Garden City, New York 11530
Phone: area code 516—294-0400

Newark Area Office
U.S. Department of Labor—OSHA
970 Broad Street—Room 1435C
Newark, New Jersey 07102
Phone: area code 201—645-5930

Newark Field Operations Office
U.S. Department of Labor—OSHA
970 Broad Street—Room 1435C
Newark, New Jersey 07102
Phone: area code 201—645-5930

New Brunswick Area Office
U.S. Department of Labor—OSHA
Bldg. T3, Belle Mead GSA Depot
Belle Mead, New Jersey 08520
Phone: area code 201—359-2777

New York Area Office
U.S. Department of Labor—OSHA
90 Church Street—Room 1405
New York, New York 10007
Phone: area code 212—264-9840

New York Field Operations Office
U.S. Department of Labor—OSHA
90 Church Street—Room 1405
New York, New York 10007
Phone: area code 212—264-9840

Puerto Rico Area Office
U.S. Department of Labor—OSHA
605 Condado Avenue—Room 328
Santurce, Puerto Rico 00907
Phone: area code 809—724-1059

Syracuse Area Office
U.S. Department of Labor—OSHA
700 East Water Street
Syracuse, New York 13210
Phone: area code 315—473-2700

Syracuse Field Operations Office
U.S. Department of Labor—OSHA
700 East Water Street—Room 203
Syracuse, New York 13210
Phone: area code 315—473-2700

White Plains Field Operations Office
U.S. Department of Labor—OSHA
200 Mamaroneck Avenue
White Plains, New York 10601
Phone: area code 914—761-4250

REGION III: DELAWARE, DISTRICT OF COLUMBIA, MARYLAND, PENNSYLVANIA VIRGINIA, WEST VIRGINIA

Regional Office
U.S. Department of Labor—OSHA
Gateway Building—Suite 15220
3535 Market Street
Philadelphia, Pennsylvania 19104
Phone: area code 215—597-1201

Baltimore Area Office
U.S. Department of Labor—OSHA
Federal Building—Room 1110A
Charles Center—31 Hopkins Plaza
Baltimore, Maryland 21201
Phone: area code 301—962-2840

Charleston Area Office
U.S. Department of Labor—OSHA
Charleston National Plaza—
Room 1726
700 Virginia Street
Charleston, West Virginia 25301
Phone: area code 304—343-1420

Falls Church Field Station
U.S. Department of Labor—OSHA
Falls Church Office Building—
Room 107
900 South Washington Street
Falls Church, Virginia 22046
Phone: area code 703—557-1330

Lancaster Field Station
U.S. Department of Labor—OSHA
3 Century Plaza—Room B
Duke and East King Streets
Lancaster, Pennsylvania 17602
Phone: area code 717—
394-9573/4

Lewistown Field Station
U.S. Department of Labor—OSHA
Post Office Building—Room 201
100 West Market Street
Lewistown, Pennsylvania 17044
Phone: area code 717—242-1441

Meadville Field Station
U.S. Department of Labor—OSHA
Burneson Building
933 Park Avenue
Meadville, Pennsylvania 16335
Phone: area code 814—724-8031

Norfolk District Office
U.S. Department of Labor—OSHA
Stanwick Building—Room 111
3661 Virginia Beach Boulevard
Norfolk, Virginia 23502
Phone: area code 804—441-6381

Philadelphia Area Office
U.S. Department of Labor—OSHA
Room 4256
William J. Green, Jr., Federal Building
600 Arch Street
Philadelphia, Pennsylvania 19106
Phone: area code 215—597-4955

Pittsburgh Area Office
U.S. Department of Labor—OSHA
Jonnet Building—Room 802
4099 William Penn Highway
Monroeville, Pennsylvania 15146
Phone: area code 412—644-2905

Richmond Area Office
U.S. Department of Labor—OSHA
Room 8015 Federal Building (P.O. Box 10186)
400 North Eighth Street
Richmond, Virginia 23240
Phone: area code 804—782-2864/5

Roanoke Field Station
U.S. Department of Labor—OSHA
Carlton Terrace Building—Suite 114
920 South Jefferson Street
Roanoke, Virginia 24011
Phone: area code 703—343-6271

Washington, D.C., Area Office
U.S. Department of Labor—OSHA
Room LL2
Railway Labor Building
400 First Street N.W.
Washington, D.C. 20210
Phone: area code 202—961-5132

Wheeling Field Station
U.S. Department of Labor—OSHA
U.S. Courthouse—Room 411
Chapline and Twelfth Streets
Wheeling, West Virginia 26003
Phone: area code 304—232-1062/3

Wilkes-Barre Field Station
U.S. Department of Labor—OSHA
IBE Building—Room 701
69 Public Square
Wilkes-Barre, Pennsylvania 18701
Phone: area code 717—825-6538

Wilmington Field Station
U.S. Department of Labor—OSHA
Federal Office Building—Room 3005
844 King Street
Wilmington, Delaware 19801
Phone: area code 302—571-6115

***REGION IV**: ALABAMA, FLORIDA, GEORGIA, KENTUCKY, MISSISSIPPI, NORTH CAROLINA, SOUTH CAROLINA, TENNESSEE*

Regional Office
U.S. Department of Labor—OSHA
1375 Peachtree Street, N.E.—Suite 587
Atlanta, Georgia 30309
Phone: area code 404—526-3573/4 or 526-2281/2 or 526-2283/4 or 526-2285/6

Anniston Field Station
U.S. Department of Labor—OSHA
1129 Noble Street—Room M104
Anniston, Alabama 36201
Phone: area code 205—237-4212

Atlanta Area Office
U.S. Department of Labor—OSHA
Building 10—Suite 33
La Vista Perimeter Office Park
Tucker, Georgia 30084
Phone: area code 404—939-8987

Birmingham Area Office
U.S. Department of Labor—OSHA
Todd Mall
2047 Canyon Road
Birmingham, Alabama 35216
Phone: area code 205—822-7100

Charleston Field Station
U.S. Department of Labor—OSHA
334 Meeting Street
Federal Building—6th Floor
Charleston, South Carolina 29403
Phone: area code 803—577-2423

Chattanooga Field Station
U.S. Department of Labor—OSHA
Eastgate Shopping Center
Building 6300—Suite 7003
Chattanooga, Tennessee 37411
Phone: area code 615—266-3273

Columbia Area Office
U.S. Department of Labor—OSHA
1710 Gervais Street—Room 205
Columbia, South Carolina 29201
Phone: area code 803—765-5904

Fort Lauderdale Area Office
U.S. Department of Labor—OSHA
Bridge Building—Room 204
3200 East Oakland Park Boulevard
Fort Lauderdale, Florida 33308
Phone: area code 305—735-6600, extension 331

Huntsville Field Station
U.S. Department of Labor—OSHA
P.O. Box 4347
Huntsville, Alabama 35802
Phone: area code 205—883-1304

Jackson Area Office
U.S. Department of Labor—OSHA
57601-55 North Frontage Road East
Jackson, Mississippi 39211
Phone: area code 601—969-4606

Jacksonville Area Office
U.S. Department of Labor—OSHA
Art Museum Plaza—Suite 4
2809 Art Museum Drive
Jacksonville, Florida 32207
Phone: area code 904—791-2895

Louisville Area Office
U.S. Department of Labor—OSHA
600 Federal Place
Suite 554-E
Louisville, Kentucky 40202
Phone: area code 502—582-6111/2

Macon Area Office
U.S. Department of Labor—OSHA
Riverside Plaza Shopping Center
2720 Riverside Drive
Macon, Georgia 31204
Phone: area code 912—746-5143

Memphis Field Station
U.S. Department of Labor—OSHA
2859 Churchill
Memphis, Tennessee 38118
Phone: area code 901—363-1584

Mobile Area Office
U.S. Department of Labor—OSHA
Commerce Building—Room 600
118 North Royal Street
Mobile, Alabama, 36602
Phone: area code 205—690-2131

Montgomery Field Station
U.S. Department of Labor—OSHA
474 South Court
Montgomery, Alabama 36103
Phone: area code 205—832-7159

Nashville Area Office
U.S. Department of Labor—OSHA
Suite 302
1600 Hayes Street
Nashville, Tennessee 37203
Phone: area code 615—749-5313

Panama City Field Station
U.S. Department of Labor—OSHA
P.O. Box 756
Lynn Haven, Florida 32444
Phone: area code 904—265-3408

Raleigh Area Office
U.S. Department of Labor—OSHA
Federal Office Building—
Room 406
310 New Bern Avenue
Raleigh, North Carolina 27601
Phone: area code 919—755-4770

Savannah Area Office
U.S. Department of Labor—OSHA
Enterprise Building—Suite 204
6605 Abercon Street
Savannah, Georgia 31405
Phone: area code 912—354-0733

Tallahassee Field Station
U.S. Department of Labor—OSHA
Kozerama Building
1300 Executive Center Drive
Tallahassee, Florida 32301
Phone: area code 904—877-3215

Tampa Area Office
U.S. Department of Labor—OSHA
650 Cleveland Street—Room 44
Clearwater, Florida 33515
Phone: area code 813—466-1769
or 813—447-6476/77

Wilmington Field Station
U.S. Department of Labor—OSHA
First Union National Bank Building
201 North Front Street
Wilmington, North Carolina 28401
Phone: area code 919—791-6430

Winfield Field Station
U.S. Department of Labor—OSHA
Route 2, Box 4
Winfield, Alabama 35594
Phone: area code 205—487-6668

REGION V: *ILLINOIS, INDIANA, MICHIGAN, MINNESOTA, OHIO, WISCONSIN*

Regional Office
U.S. Department of Labor—OSHA
300 South Wacker Drive—
Room 1201
Chicago, Illinois 60606
Phone: area code 312—
353-4716/7

Chicago Area Office
U.S. Department of Labor—OSHA
230 South Dearborn—10th Floor
Chicago, Illinois 60604
Phone: area code 312—353-1390

Cincinnati Area Office
U.S. Department of Labor—OSHA
Federal Office Building—
Room 4028
550 Main Street
Cincinnati, Ohio 45202
Phone: area code 513—684-2354

Cleveland Area Office
U.S. Department of Labor—OSHA
Federal Office Building—
Room 847
1240 East Ninth Street
Cleveland, Ohio 44199
Phone: area code 216—522-3818

Columbus Area Office
U.S. Department of Labor—OSHA
Room 109
360 South Third Street
Columbus, Ohio 43215
Phone: area code 614—469-5582

Detroit Area Office
U.S. Department of Labor—OSHA
Michigan Theatre Building—
Room 626
220 Bagley Avenue
Detroit, Michigan 48226
Phone: area code 313—226-6720

Indianapolis Area Office
U.S. Department of Labor—OSHA
U.S. Post Office and
Courthouse—Room 423
46 East Ohio Street
Indianapolis, Indiana 46202
Phone: area code 317—633-7384

Milwaukee Area Office
U.S. Department of Labor—OSHA
Clark Building—Room 400
633 West Wisconsin Avenue
Milwaukee, Wisconsin 53203
Phone: area code 414—
224-3315/6

Minneapolis Area Office
U.S. Department of Labor—OSHA
110 South Fourth Street—
Room 437
Minneapolis, Minnesota 55401
Phone: area code 612—725-2571

Toledo Area Office
U.S. Department of Labor—OSHA
Federal Office Building—Room 734
234 North Summit Street
Toledo, Ohio 43604
Phone: area code 419—259-7542

***REGION VI:** ARKANSAS, LOUISIANA, NEW MEXICO, OKLAHOMA, TEXAS*

Regional Office
U.S. Department of Labor—OSHA
Texaco Building—7th Floor
1512 Commerce Street
Dallas, Texas 75201
Phone: area code 214—
749-2477/8/9

Albuquerque Area Office
U.S. Department of Labor—OSHA
Federal Building—Room 302
421 Gold Avenue S.W.
P.O. Box 1428
Albuquerque, New Mexico 87103
Phone: area code 505—766-3411

Corpus Christi District Office
U.S. Department of Labor—OSHA
Suite 1322
600 Leopard Street
Corpus Christi, Texas 78401
Phone: area code 512—888-5416

Dallas Area Office
U.S. Department of Labor—OSHA
Adolphus Tower—Suite 1820
1412 Main Street
Dallas, Texas 75202
Phone: area code 214—
749-1786/7/8

El Paso Field Station
U.S. Department of Labor—OSHA
Room 3
1515 Airway Boulevard
El Paso, Texas 79925
Phone: area code 915—543-7828

Fort Worth Field Station
U.S. Department of Labor—OSHA
Room 8A36
819 Taylor Street
Fort Worth, Texas 76102
Phone: area code 817—334-2881

Houston Area Office
U.S. Department of Labor—OSHA
Room 2118
2320 La Branch Street
Houston, Texas 77004
Phone: area code 713—226-5431

Little Rock Area Office
U.S. Department of Labor—OSHA
Donaghey Building—Room 526
103 East Seventh Street
Little Rock, Arkansas 72201
Phone: area code 501—378-6192

Lubbock Area Office
U.S. Department of Labor—OSHA
Federal Building—Room 421
1205 Texas Avenue
Lubbock, Texas 79401
Phone: area code 806—762-7681

New Orleans Area Office
U.S. Department of Labor—OSHA
Room 202
546 Carondelet Street
New Orleans, Louisiana 70130
Phone: area code 504—527-2451/2

Oklahoma City Field Station
U.S. Department of Labor—OSHA
Suite 408
50 Penn Place
Oklahoma City, Oklahoma 73118
Phone: area code 405—231-5351

San Antonio Area Office
U.S. Department of Labor—OSHA
Room 122
1015 Jackson Keller Road
San Antonio, Texas 78213
Phone: area code 512—225-4569

Shreveport Field Station
U.S. Department of Labor—OSHA
New Federal Office Building—Room 8A09
500 Fannin Street
Shreveport, Louisiana 71101
Phone: area code 318—226-5360

Tulsa Area Office
U.S. Department of Labor—OSHA
Petroleum Building—Room 512
420 South Boulder Avenue
Tulsa, Oklahoma 74103
Phone: area code 918—581-7676

***REGION VII:** IOWA, KANSAS, MISSOURI, NEBRASKA*

Regional Office
U.S. Department of Labor—OSHA
Room 3000
911 Walnut Street
Kansas City, Missouri 64106
Phone: area code 816—374-5861

Des Moines Area Office
U.S. Department of Labor—OSHA
Room 643
210 Walnut Street
Des Moines, Iowa 50309
Phone: area code 515—284-4794

Kansas City Area Office
U.S. Department of Labor—OSHA
Room 1100
1627 Main Street
Kansas City, Missouri 64108
Phone: area code 816—374-2756

North Platte Area Office
U.S. Department of Labor—OSHA
113 West Sixth Street
North Platte, Nebraska 69101
Phone: area code 308—534-9450

Omaha Area Office
U.S. Department of Labor—OSHA
Room 803—Harney and Sixteenth Streets
City National Bank Building
Omaha, Nebraska 68102
Phone: area code 402—221-3276

St. Louis Area Office
U.S. Department of Labor—OSHA
Room 554
210 North Twelfth Boulevard
St. Louis, Missouri 63101
Phone: area code 314—622-5461

Wichita Area Office
U.S. Department of Labor—OSHA
Petroleum Building—Suite 312
221 South Broadway Street
Wichita, Kansas 67202
Phone: area code 316—267-6311, extension 644

REGION VIII: *COLORADO, MONTANA, NORTH DAKOTA, SOUTH DAKOTA, UTAH, WYOMING*

Regional Office
U.S. Department of Labor—OSHA
Federal Building—Room 15010
1961 Stout Street
Denver, Colorado 80202
Phone: area code 303—837-3883

Billings Area Office
U.S. Department of Labor—OSHA
Petroleum Building—Suite 525
2812 First Avenue North
Billings, Montana 59101
Phone: area code 406—245-6711, extensions 6640/9

Bismarck Area Office
U.S. Department of Labor—OSHA
Russell Building
Highway, 83 North
Route 1
Bismarck, North Dakota 58501
Phone: area code 701—255-4011, extension 521

Lakewood Area Office
U.S. Department of Labor—OSHA
Squire Plaza Building
8527 West Colfax Avenue
Lakewood, Colorado 80215
Phone: area code 303—234-4471

Salt Lake City Area Office
U.S. Department of Labor—OSHA
U.S. Post Office Building—Room 452
350 South Main Street
Salt Lake City, Utah 84101
Phone: area code 801—524-5080

Sioux Falls Area Office
U.S. Department of Labor—OSHA
Court House Plaza Building—Room 408
300 North Dakota Avenue
Sioux Falls, South Dakota 57102
Phone: area code 605—336-2980, extension 425

REGION IX: *ARIZONA, CALIFORNIA, HAWAII, NEVADA, AMERICAN SAMOA, GUAM, TRUST TERRITORY OF THE PACIFIC ISLANDS*

Regional Office
U.S. Department of Labor—OSHA
9470 Federal Building
450 Golden Gate Avenue—P.O. Box 36017
San Francisco, California 94102
Phone: area code 415—556-0586

Carson City Area Office
U.S. Department of Labor—OSHA
1100 East William Street
Carson City, Nevada 89701
Phone: area code 702—883-1226

Fresno Field Station
U.S. Department of Labor—OSHA
Suite C
426 North Abby Street
Fresno, California 93701
Phone: area code 209—487-5454

Honolulu Area Office
U.S. Department of Labor—OSHA
333 Queen Street—Suite 505
Honolulu, Hawaii 96813
Phone: area code 808—546-3157

Las Vegas Field Station
U.S. Department of Labor—OSHA
Room 4-627
300 Las Vegas Boulevard South
Las Vegas, Nevada 89101
Phone: area code 702—385-6750

Long Beach Area Office
U.S. Department of Labor—OSHA
Hartwell Building—Room 401
19 Pine Avenue
Long Beach, California 90802
Phone: area code 213—432-3434

Phoenix Area Office
U.S. Department of Labor—OSHA
Amerco Towers—Suite 318
2721 North Central Avenue
Phoenix, Arizona 85004
Phone: area code 602—261-4858

Sacramento Field Station
U.S. Department of Labor—OSHA
Room 1409
2800 Cottage Way
Sacramento, California 95825
Phone: area code 916—484-4363

San Diego Field Station
U.S. Department of Labor—OSHA
Suite 308
770 B Street
San Diego, California 92101
Phone: area code 714—293-5396

San Francisco Area Office
U.S. Department of Labor—OSHA
Room 1706
100 McAllister Street
San Francisco, California 94102
Phone: area code 415—556-0536

Tucson Field Station
U.S. Department of Labor—OSHA
Room 204
155 East Alameda Street
Tucson, Arizona 85701
Phone: area code 602—792-6286

REGION X: ALASKA, IDAHO, OREGON, WASHINGTON

Regional Office
U.S. Department of Labor—OSHA
Room 6048, Federal Office Building
909 First Avenue
Seattle, Washington 98174
Phone: area code 206—442-5930

Anchorage Area Office
U.S. Department of Labor—OSHA
Federal Building—Room 227
605 West Fourth Avenue
Anchorage, Alaska 99501
Phone: area code 907—265-5341

Bellevue Area Office
U.S. Department of Labor—OSHA
121 - 107th Street, N.E.
Bellevue, Washington 98004
Phone: area code 206—442-7520

Boise Area Office
U.S. Department of Labor—OSHA
228 Idaho Building
216 North Eighth Street
Boise, Idaho 83702
Phone: area code 208—342-2622

Pocatello Field Station
U.S. Department of Labor—OSHA
Room B17
Federal Building
U.S. Courthouse
150 South Arthur Street
Pocatello, Idaho 83201
Phone: area code 208—233-6374

Portland Area Office
U.S. Department of Labor—OSHA
Pittock Block—Room 526
921 Southwest Washington Street
Portland, Oregon 97205
Phone: area code 503—221-2251

Spokane Field Station
U.S. Department of Labor—OSHA
Room 410—U.S. Post Office
West 904 Riverside Avenue
Spokane, Washington 99201
Phone: area code 509—456-2598